Carlos Kiangebeni Zeye

The Wonders of Moringa

Carlos Kiangebeni Zeye

The Wonders of Moringa

Application of Moringa Oleifera Lam to the treatment of domestic sewage mixed with industrial effluent

ScienciaScripts

Imprint

Any brand names and product names mentioned in this book are subject to trademark, brand or patent protection and are trademarks or registered trademarks of their respective holders. The use of brand names, product names, common names, trade names, product descriptions etc. even without a particular marking in this work is in no way to be construed to mean that such names may be regarded as unrestricted in respect of trademark and brand protection legislation and could thus be used by anyone.

Cover image: www.ingimage.com

This book is a translation from the original published under ISBN 978-620-6-75800-6.

Publisher:
Sciencia Scripts
is a trademark of
Dodo Books Indian Ocean Ltd. and OmniScriptum S.R.L publishing group

120 High Road, East Finchley, London, N2 9ED, United Kingdom
Str. Armeneasca 28/1, office 1, Chisinau MD-2012, Republic of Moldova, Europe
Printed at: see last page
ISBN: 978-620-7-79788-2

Contents

ACKNOWLEDGEMENTS

To God for the gift of life and for always being with me.

To my wife: Holanda Yamba and children: Efraim Carlos, Moises Carlos, Barzilai Carlos, Jasobeam Carlos and Rode Zeye for their love and complicity in always believing in me. To all my family for their constant encouragement and support.

To my supervisor Prof^a Dr Anabela Alexandre Leitao, for her availability, patience, openness in transmitting knowledge, suitability and character in providing guidance and encouragement, which made it possible to complete this work.

To Master Rosalina Serrao for her unstinting supervision of the laboratory experiments and for her patience in helping with the work, on certain occasions having to leave the laboratory very late at night.

To Fernando Jose for his technical support, his very pertinent contributions throughout the realisation of the experiments.

To Master Josefa Will for her technical support and participation in the conditioning process because she agreed to leave the laboratory very late; to Technicians Arao Chipate and Timoteo for their skilful participation at various times during the work.

The team responsible for managing the wastewater treatment plants in Luanda: Engineers Cfcero and Vftor, and the operators of the Viana/Luanda EEZ wastewater treatment plant for their availability and supply of the samples used in this work.

To Prof Dr Pedro Guilherme for encouraging me to study and for his moral support in carrying out this study.

To Prof Dr Aderito Cunha for his spirit of openness, motivation and encouragement.

To my colleagues in the service: Rosa Lourengo, Cesaltina Balo, Jose Munguengue, Joao Carlos, Joaquim Cordeiro, Lucio Boamorte, Kuta Filipe, Sandra Calundungu, Tshiluka Itwev, Jeronimo Diogo, Aristide liassoka, Mayi Mbemba... for the moral and material support they encouraged to carry out the work.

To Kengana Albertina for her encouragement, moral and financial support in the most critical moments of this work.

To Apostle Leopold Mutombo Kalombo for his spiritual support and impact. To Jose Kamongua and Filipe Kavula for their moral and financial support at critical times.

To all the people who directly or indirectly collaborated in the execution of this work. Thank you very much!

SUMMARY

The current reality of wastewater treatment in the country leads to the analysis of low-cost treatments using simple technologies to minimise operational difficulties. There are many techniques used in water treatment that employ natural agents to remove turbidity, organic matter and metals. In this context, the aim of this study was to separately evaluate the efficiency of coagulation-flocculation techniques in removing turbidity and adsorption in removing dissolved organic matter and metals from wastewater treated at the ZEE WWTP. For this purpose, parts of Moringa oleifera Lam were used, namely ground seeds with and without husks and their aqueous and saline extracts in the coagulation-flocculation tests and seed pods and husks in the adsorption tests. In the coagulation-flocculation tests, the best results were achieved with the ground seed without the shell, obtaining a turbidity removal of 97% (final turbidity of 10.7 NTU) with 0.12 g/L of the material at a pH of 12. In the adsorption tests, the best adsorbent for organic matter was the shell because it had the highest adsorption capacity and a maximum organic matter removal of 90% obtained with 0.3 g of the adsorbent at 20 °C. This adsorbent was also tested for Pb adsorption and a maximum removal of 78.6% was obtained with 0.5 g of the adsorbent at 20 °C. The results obtained suggest a promising combined process for treating a mixture of domestic sewage and industrial effluent consisting of coagulation-flocculation followed by adsorption.

Keywords: Domestic sewage, industrial effluent, treatment, *Moringa oleifera* Lam.

1 INTRODUCTION

1.1 Framework and justification for the study

Worldwide, the growth of industrial activities has intensified and has caused environmental pollution that leads to the deterioration of ecosystems, often caused by man himself (Lenardao, Freitag, Dabdoub, Batista, & Silveira, 2003). Unchecked population growth increases problems of anthropogenic origin that easily lead to altered soil, air and water quality (Kunz, Zamora, Moras, & Duran, 2002).

In Angola, especially in the city of Luanda, the growth of urbanisation, industrialisation and anthropogenic activity contributes greatly to the production of sewage, which in most cases is directly discharged into receiving environments without prior treatment. Therefore, the failure to treat wastewater of domestic or industrial origin and its discharge into the receiving environment can lead to a number of problems, such as the pollution of surface and groundwater, used as a source of drinking water, and the consequent risk of spreading diseases.

Today, it is becoming imperative to investigate simple, low-cost processes for treating wastewater produced as a result of domestic and industrial activities.

In Angola there is a set of normative principles, set out in Presidential Decree 261/11 (2011), which regulates the emission/discharge of wastewater into water or soil by a facility that requires a licence to be issued by the Ministry of the Environment, which sets out the discharge standards for mitigating or preventing the applicable damage. In other words, the decree establishes the limits for the emission/discharge of each type of pollutant from a sewer into a watercourse, given that the self-depuration capacity of the receiving environments is increasingly limited.

According to Almeida (2018), Moringa is used in various areas of research or in various industrial sectors, but its main application is in the field of Chemical Engineering, namely in water treatment as a natural coagulant. Nowadays, it seems to be an alternative for improving the living conditions of the rural population and preserving the environment, mainly due to its low cost and low toxicity. Silva and Matos (2010) reported that in the context of using low-cost environmental technologies based on natural coagulants and flocculants produced in the environment itself, the seeds and pods of *Moringa oleifera* Lam proved to be efficient in sewage treatment.

Normally, the application of natural coagulants or flocculants has shown advantages over chemical ones, especially in terms of biodegradability, low toxicity and low sludge production rates.

The lack of alternatives for the treatment of domestic sewage and industrial effluents in the city of Luanda is a serious environmental problem with a strong negative impact on public health. This problem is aggravated, on the one hand, by the high population density resulting from demographic expansion and rural exodus, and, on the other hand, by the intensification of commercial activity due to the restructuring of the industrial sector and its production chain.

These factors increase not only the need for water consumption, but also the production of discharges in the form of effluents. The majority of existing sewage channels and pipes end in an untreated watercourse. The high density of rudimentary constructions near rivers and lakes poses a risk to people's health. Therefore, the urban environment and natural receptors such as rivers are becoming increasingly fragile and require greater protection from pollution. In addition, this pollution affects bathing waters and fishery products. These environmental, economic and public health considerations justify the preparation of this study as a future contribution to the different alternatives for reducing environmental liabilities, showing how *Moringa oleifera* Lam can be applied to the treatment of domestic sewage mixed with industrial effluent.

This research work contributes to the development of proposals for low-cost treatment of domestic sewage mixed with industrial effluent, since *Moringa oleifera* Lam has coagulant power for suspended particles and adsorptive capacity for organic matter and metals in solution. The treatment will be possible through natural adsorption and coagulation-flocculation mechanisms, using seed husks and pods as adsorbents and ground seeds and seed extracts as coagulants, thus saving resources on the import of chemical coagulants used in sewage treatment in our country.

The work carried out consisted of testing the purifying capacity of *Moringa oleifera* Lam seeds and pods in order to verify their viability as a coagulant, flocculant and adsorbent material.

To carry out this work, the laboratory tests were carried out at the Laboratory of Separation Engineering, Chemical Reaction and Environment (LESRA), associated with the Faculty of Engineering of the Agostinho Neto University. The sewage samples to be used were collected from the Special Economic Zone (ZEE) Wastewater Treatment Plant in Viana - Luanda, and the Moringa seeds and pods were collected from Bairro da Fubu, Camama Commune, Kilamba-Kiaxi District, Belas Municipality.

1.2 Study objectives

1.2.1 General objective

- To evaluate the possibility of using *Moringa oleifera* Lam seeds and pods from Angola as a coagulant and adsorbent to treat industrial wastewater mixed with domestic sewage.

1.2.2 Specific objectives

•   Evaluate the removal of turbidity and organic matter from industrial wastewater mixed with domestic sewage by coagulation and flocculation using seeds as they are and protein extracted from the seeds.

•   Quantify the adsorption capacity of seed husks and pods for removing organic matter and metals present in a sample of industrial wastewater mixed with domestic sewage.

•   Determine the adsorption kinetics of organic matter and metals through experiments in closed adsorbers.

1.3 Dissertation structure

The dissertation is structured in five chapters:

Chapter 1 - Introduction. This chapter briefly discusses the reasons for treating domestic sewage mixed with industrial effluent from the Viana EEZ WWTP in Luanda (Angola) and presents the justification for this study. This chapter also includes the general and specific objectives of the work and the structure of the dissertation.

Chapter 2 - Literature review. This chapter presents the bibliographical research carried out on the processes of treating domestic sewage mixed with industrial effluent, on *Moringa oleifera* Lam and its intensive application in the treatment of domestic and industrial sewage as a coagulant-flocculant and natural adsorbent and the results of work already carried out.

Chapter 3 - Methodology. This chapter presents in detail the materials and methods used to carry out this work. The characteristics of domestic sewage mixed with industrial effluent, the location of the sampling site and biological material, the preparation of *Moringa oleifera* Lam as a coagulant-flocculant and adsorbent are described, the experimental protocol related to the application of *Moringa oleifera* Lam as a coagulant-flocculant, the adsorption kinetics and equiKbrium tests carried out, the parameters measured, their methods of analysis and the materials used.

Chapter 4 - Results and Discussion. This chapter contains the experimental results of the study and their treatment, the interpretation of the results obtained, relating them to known facts and scientific principles and concepts. These results are compared with those of other works in the same field, explaining and justifying the exceptions found in the results of this work and showing the implications of this work for the field of study.

Chapter 5 - Conclusions. This chapter summarises the relevant results obtained and proposes the continuation of the studies begun in this research project.

2 LITERATURE REVIEW

From the 1960s onwards, socio-economic transformations began to emerge, fostering industrial growth and the appeal of unusual consumption, as well as demographic growth and migration, which led to a very marked process of urbanisation.

These transformations have actually been harmful to the environment, especially in urban centres, where pollution and its contaminating forms have become more evident.

Pollution has hit water resources hardest, followed by air and soil.

In recent decades, public awareness of water pollution problems has grown considerably, increasing the number of increasingly stringent regulations on wastewater discharge (Zeng *et al.*, 2007).

In Angola, water management with the aim of protecting the aquatic environment and improving its quality, depending on its main uses, is regulated by Presidential Decree No. 261/11 (2011), which establishes water quality standards and criteria for different uses, including the production of drinking water. The provisions of this decree apply to inland waters, both surface and groundwater, as well as waters for aquaculture, livestock farming, agricultural irrigation and bathing. In addition, this decree-law regulates the rules for controlling the discharge of waste water into national aquatic bodies and the soil, with a view to preserving the quality of the aquatic environment and protecting public health.

According to Giordano (2020), the characteristics of industrial effluents are inherent to the composition of the raw materials, the water supply and the industrial process.

Given the toxicity of industrial effluent compounds, pre-treatment must be carried out to remove pollutants that may make biological treatment unfeasible. This pre-treatment depends on the quality and quantity of the effluent, and many techniques are used to remove pollutants.

2.1 Wastewater pollutants

In wastewater, pollutants are dissolved or suspended (coarse, fine and colloidal). The separation of these pollutants along the treatment chain gives rise to the treated effluent. Pollutants also include solid waste (silt, sand and sludge), organic matter, nutrients (nitrogen and phosphorus) and gases such as hydrogen gas, carbon dioxide, methane and other gases in smaller volumes.

2.1.1 Metals

Metals are the largest group of chemical elements. By definition, they are good conductors of electricity and their electrical conductivity decreases with temperature. This leads us to differentiate between non-metals and metalloids (B, Si, Ge, As and Te). In other words, the former are not good conductors and the latter are weak conductors of electricity and their electrical conductivity increases with temperature (Metals handout, 2009).

According to Duarte and Pasqual (2000), the term heavy metal in principle designates a metal that occurs in natural systems in small concentrations and has a density equal to or greater than 5 g/cm^3 .

Heavy metals are chemical elements that have an atomic number greater than 22, for example: Fe (26); Ni (28); Cu (29); Zn (30); Cd (48) and Pb (82). They can also be defined by their unique property of precipitating as sulphides (Perpetuo, 2020).

Mota *et al.* (2005) and Mello *et al.* (2005) cited by Santos (2012) defined "trace metals", in particular heavy metals, as important elements in natural aquatic systems, with total concentrations in the environment of less than $10\text{-}8\ M$ and with concentrations of their free ions as low as 10-^{12} M. Some metals such as zinc, copper, selenium, bismuth, manganese and nickel can be considered either essential or toxic elements, depending on their concentrations and the organisms considered.

There are three groups of metals: essential metals, toxic metals and indifferent metals (Monteiro, 2009).

- The essentials have essential functions in living matter: Na, K, Mg, Ca, V, Mn, Fe, Co, Ni, Cu, Zn, Mo and W, of which the first four are building blocks in the intercellular environment, are called macronutrients and have a high concentration. The others, such as Zn, Mo, Co and Ni, are necessary at low concentrations for the proper performance of metabolic activities, but become toxic at high concentrations and are called micronutrients.

- The toxic or non-essential ones have no known metabolic function: Ag, Cd, Al, Bi, Pb, Ti, Hg, Au and Sn.

- The indifferent or non-sperm-functioning ones are Sr, Cs, Rb and various transition metals, which can be accumulated in the cells by non-sperm physico-chemical interactions or by sperm transport mechanisms.

Industrial effluents such as those generated in metal extraction plants, paint and pigment industries, especially electroplating plants, which are scattered in large numbers on the outskirts of large cities, contain chemical contaminants that pollute water and have unfavourable effects on health.

Heavy metals can be present in effluents from chemical industries, such as those that formulate organic compounds and inorganic elements and compounds, leather, fur and similar product industries, iron and steel industries, laundries and the oil industry (Perpetuo, 2020).

The preferential pathways by which metals are transported in water depend on various physical, chemical and biological factors. In general, waters that receive effluents containing heavy metals have high concentrations of these in the bottom sediment.

From all these sources, living beings assimilate metals, accumulate them and transfer them along trophic chains.

According to Monteiro (2009), the toxic effects that manifest themselves throughout the body are associated with their dose and chemical form, and can affect various organs, altering biochemical processes, organelles and cell membranes.

As far as the environment and public health are concerned, the most common sources of metal contamination are fertilisers, pesticides, vehicle emissions, mining, coal and oil combustion, smelting, the refining of pollutants and the incineration of urban and industrial waste (Monteiro, 2009).

2.1.1.1 Lead (Pb)

It is a bluish-grey metal, shiny, odourless, easily malleable, ductile, insoluble in organic solvents, a weak conductor of electricity, resistant to corrosion, but becomes opaque when exposed to air (Monteiro, 2009). The sources are mostly natural, such as volcanic emissions, geochemical weathering, water fogs and geological sources in igneous and metamorphic rocks (Kreush, 2005).

2.1.1.2 Zinc (Zn)

It is a bluish-white metal with a compact hexagonal crystalline form, found in nature mainly in the form of sulphides, associated with lead, copper, silver and iron. Zinc sulphide undergoes many transformations in the oxidation zone, forming oxides, carbonates and silicates. These transformations or mineralisations occur in limestone rocks.

2.1.1.3 Nickel (Ni)

It is a silver-white, ductile, malleable metal. Its important sources are mines in the form of sulphide, millerite and pentlandite, and it is associated with other metallic sulphides in basal rocks, often accompanied by copper and cobalt. These contribute more than 90% of the nickel extracted (Petroni & Pires, 2000).

2.1.1.4 Cadmium

It is considered a heavy metal, one of the most harmful to health due to its high toxicity (Dias, 2001). It is widely used to coat materials, as a paint pigment and in the plastics industry. Cd can be added to the soil through urban or industrial waste, sewage sludge and phosphate fertilisers (Dias, 2001).

2.1.1.5 Chromium

Chromium is widely used in industry, especially in electroplating, where chrome plating is one of the most common handle coatings. It can occur as a contaminant in water subject to tannery effluent leaks and in the circulation of cooling water, where it is used to control corrosion. The hexavalent form is more toxic than the trivalent. It produces corrosive effects on the digestive system and nephritis (Perpetuo, 2020).

2.1.2 Organic matter

According to Monte *et al.* (2016), wastewater can contain dissolved organic and inorganic substances suspended in the water from surface or groundwater, which is the source of raw water for human consumption. This water also contains substances from other sources. However, wastewater is a complex mixture of dissolved and suspended substances, with a considerable number of microorganisms of various types, most of which are of faecal origin and some of which are pathogenic.

Organic compounds are those that contain carbon atoms, with the exception of compounds in the CO_2 family such as carbonates and hydrogen carbonates, cyanides and other compounds.

Examples of these substances that are present in wastewater are: petroleum products, rubber, plastics and synthetic fibres, antibiotics and vitamins, agricultural fertilisers, pesticides, sugars, dyes, paints, etc.

The vast majority of organic compounds are biodegradable, meaning that the process of transforming organic compounds into mineral or inorganic compounds is carried out by the action of living beings, particularly microorganisms.

Examples of biodegradable compounds found in wastewater are proteins, carbohydrates and carbohydrates (including fats and oils) and they account for 40 to 60 per cent, 25 to 50 per cent and around 10 per cent respectively (Monte *et al.*, 2016).

2.1.3 Solids

Wastewater is made up of a mixture of dissolved and suspended substances or solids and can be loaded with various types of microorganisms, many of faecal origin and some pathogenic.

2.1.3.1 Total Solids (ST)

They are present in wastewater and are one of its most important physical characteristics, comprising organic and inorganic substances in solution and suspension. These total solids are quantified by weighing the dry residue obtained after evaporation of the sample water.

In general, Total Solids of an organic nature are called Total Volatile Solids (TVS) because they volatilise after calcination when subjected to temperatures of 550 ± 50 °C in a muffle furnace. The dry residue obtained constitutes the inorganic part of the solids present in the wastewater and is called Total Fixed Solids (TFS). The determination of SVT and SFT does not make it possible to distinguish between organic and inorganic material with precision, because calcination leads to a loss of some volatile inorganic compounds.

2.1.3.2 Sedimentable Solids (SSed)

They are those in suspension that sediment easily, represent the amount of solids that sediment in an Imhoff cone over a period of 60 minutes and are expressed in ml/L.

2.1.3.3 Total Suspended Solids (TSS)

These are the ones that are retained by filtering the waste water through a filter (usually made of glass fibre) with a nominal pore size of 1.2 pm.

2.1.4 Nitrogen

In the biosphere cycle, nitrogen alternates between various forms and oxidation states as a result of different biochemical processes. In the aquatic environment, it is found in the following forms: molecular nitrogen (Nº), organic nitrogen (variable), free ammonia (NH_3), ammonium ion (NH_4), nitrite ion (NO_2) and nitrate ion (NO_3).

2.1.5 Phosphorus

In wastewater, total phosphorus is presented as phosphates in different forms: inorganic (polyphosphates and orthophosphates), organic (bound to organic compounds).

2.2 Wastewater treatment processes

Domestic sewage treatment processes can include several sequential treatment steps, involving physical, chemical and biological processes depending on the nature of the pollutants to be removed, as described below:

2.2.1 Preliminary treatment

Preliminary treatment consists of removing coarse solids, sand particles and oil and grease. The solids found in the sewage can cause clogging damage to electromechanical equipment and pipework installed downstream. In this preliminary treatment, physical removal mechanisms predominate, making it possible to remove coarse solids and grains with diameters greater than 40 mm and 6 mm in the grates and desanders, respectively.

a) Railing

Grating consists of separating coarse materials in suspension by means of grids in order to avoid blockages in the pipes when transporting the effluent, to avoid damage to electromechanical equipment, to avoid reducing the useful volume of the biological reactor occupied with biomass and consequent problems in the treatment. The spacing between the bars varies between 0.5 and 2 cm and there are two types: simple grids used to remove solids whose volume is not very large, which are cleaned manually, and mechanised grids used to remove debris (solids whose volume is less than that of the solids removed manually), which are cleaned mechanically.

b) Desander and degreaser

The desander is made up of two parallel channels that allow the sand particles inside to settle during their journey. The two channels work independently, so that when one is in operation, the other is being maintained and cleaned.

Degreasing consists of removing oils and fats or other substances less dense than water that can affect the functioning of water treatment operations. Oils and fats are easily separated using a floatation tank that has a compressed air blower at the bottom in order to accelerate the ascent of the fats that are attached to the air bubbles (Silveira, 2010).

The fat can be collected on the surface of the tank by overflow or by using a scraper to help remove it. It is separated by the difference in density between the water and the fats or oils (Silveira, 2010).

c) Sieving

Sieving is the separation of finer particles through a fine mesh in which smaller solids are retained. Sieving is used in addition to harrowing and is more efficient at removing fine solids.

d) Parshall Channel

Parshall channels are used to measure the flow of wastewater. Through the throttled and raised part of the channel, a relationship is established between the current flow and the height of the water lamina in that region for a given vertical section upstream. Channels reduce head loss and are necessary for reading flows (Silveira, 2010).

e) Equalisation tank

The main purpose of equalisation is to regulate the flow rate because the water treatment process requires a constant flow rate. If the flow rate varies too much throughout the process, this is harmful because it makes it impossible for the pH correction tank to function normally. Equalisation tanks also homogenise effluent to make it uniform. There are several reasons why they should be used, such as: minimising operational problems caused by variations in the characteristics of the effluent and shocks caused by overloading the system, improving biological treatment, diluting inhibiting substances, stabilising the pH and improving the final quality of the effluent to be treated (Silveira, 2010).

2.2.2 Primary treatment

The primary treatment stage consists of physical and chemical processes, in which the effluent passes through a sedimentation unit (primary decanter) and the heavier particles or settleable solids are deposited at the bottom. This happens after the effluent has undergone preliminary treatment to improve the removal of settleable solids.

Primary treatment alone can remove around 50 to 70 per cent of suspended solid matter (TSS) and reduce BOD_5 by 20 to 35 per cent (Leitao, 2018). Primary treatment takes place in the primary decanter, which can be rectangular or circular.

The effluent that comes from the preliminary treatment contains suspended solids in which the denser solids gradually sediment to form the crude primary sludge. This type of sludge is removed with a scraper so as not to jeopardise the effectiveness of the treatment (Von Sperling, 1996).

According to Stein (2012), the effectiveness of primary treatment can be increased with the use of coagulating agents. This treatment stage also includes flocculators which, with the addition of chemical products (flocculating agents), cause the flocs to increase. The best floc formation occurs when the effluent is gently agitated.

2.2.3 Secondary treatment

It is considered to be a biological treatment in which bacteria degrade organic matter and are also responsible for transforming some inorganic compounds such as ammonia, sulphates, phosphates and nitrates. The bacteria grow rapidly and increase their mass, which ends up flocculating and sedimenting, thus separating the water and forming sludge. This degradation operation takes place under aerobic or anaerobic conditions (Morais, 2013).

According to Morais (2013), in the aerobic process, the oxygen present in the system facilitates the oxidation of organic matter, largely into carbon dioxide and water. It should be noted that the reaction releases a lot of energy, which will be used by the microorganisms for their reproduction in the form of biomass, as well as in their endogenous metabolism. Part of the organic matter will be used for the growth of the microorganisms. The reactions of the aerobic metabolism of microorganisms are represented by two chemical equations:

Oxidation and synthesis:

$$COHNS + O_2 + Nutr \rightarrow CO_2 + NH_3 + C_5H_7NO_2 + OutroProdFin \quad (2.1)$$

Endogenous respiration:

$$C_5H_7NO_2 + 5O_2 \rightarrow 5CO_2 + 2H_2O + NH_3 + Energie \quad (2.2)$$

In the effluent, nitrogen can appear in the forms of ammonia (NH3 and $NH4^+$), nitrite (NO2) and nitrate (NO3$^-$) and as organic nitrogen.

During biological treatment, part of the organic nitrogen is converted into ammonia. This ammonia can subsequently be oxidised by special aerobic bacteria to form nitrates with a high aerobic oxygen consumption and can lead to a high alkalinity deficiency.

The complete biochemical oxidation of ammonia takes place in two phases (Morais, 2013):

The oxidation of ammonia to nitrite, a reaction mostly carried out by bacteria of the *Nitrosomonas* genus, follows the equation (2.3)

$$2NH_4^+ + 3O_2 \rightarrow 2NO_2^- + 4H^+ + 2H_2O + 550\,kJ \quad (2.3)$$

This is followed by the nitrite to nitrate conversion reaction carried out by Nitrobacter bacteria according to equation (2.4)

$$NO_2^- + {}^1/_2 O_2 \rightarrow NO_3^- + 75\,kJ \quad (2.4)$$

The chemical-biological process of converting nitrite to nitrate is qualified as exothermic and the microorganisms use this energy to assimilate carbon dioxide, thus suppressing the need for carbon for the nitrifying organisms. Nitrification is conditioned by certain factors such as the presence of oxygen and alkalinity, which must be sufficient to neutralise the H ions$^+$ produced in this process.

The nitrates from the nitrification operation are converted to nitrogen through the denitrification process. This takes place in an anoxic environment, i.e. a biological environment with little or no dissolved oxygen. This is where nitrate (NO_3^-) is reduced to nitrous oxide (N_2O) and gaseous nitrogen (N_2), where nitrate is the final electron acceptor instead of oxygen, as in the nitrification process. As a result, nitrogen in the form of N2 will be released into the atmosphere as a noble gaseous element. Denitrifying, facultative and heterotrophic bacteria will consume organic matter to obtain energy, using the oxygen in nitrates. According to Leitao (2018) after secondary treatment, 90 to 98% of suspended solid matter (TSS) is removed, 85 to 95% of BOD_5, 30 to 40% of Total Nitrogen and 30 to 45% of Total Phosphorus.

2.2.4 Tertiary treatment or thinning

Tertiary treatment is generally used to remove additional pollutants from wastewater with a view to discharging it into the receiving body and/or recirculating it back into the system.

Tertiary treatment processes are carried out according to needs and are very diverse. However, the following processes can be mentioned: filtration, chlorination or ozonation to remove bacteria, adsorption on activated carbon to remove colour, soluble organic matter and soluble metals. There are other processes involved in reducing foam and inorganic solids, such as electrolysis, reverse osmosis and ion exchange.

2.2.5 Adsorption

The adsorption process takes place in the solid-fluid interphase, where molecules from a fluid phase called the adsorbate (gas or liquid) adhere to a solid surface (adsorbent).

This movement or migration from the fluid phase to the solid phase is the fundamental property of adsorption and its origin is mainly related to an imbalance of surface forces that exist in the adsorbent.

There are two types of adsorption: chemical adsorption and physical adsorption.

In chemical adsorption or chemisorption, there is an effective exchange of electrons between the adsorbent and the adsorbate. In this process, the adsorbate binds more strongly to the surface of the solid through strong, irreversible interactions. In general, the molecules are attracted to the active regions of the surface and usually involve a single layer (Melo, 2009). In physical adsorption or physisorption, attraction occurs between the solid surface of the adsorbent and the adsorbate molecules with weak Van Der Waals-type interactions. In this case, the process can be reversed (Melo, 2009).

2.2.5.1 Adsorbents

An adsorbent is an insoluble solid surface, usually porous and with a significant surface area, suitable for adhering molecules dispersed in a liquid or gaseous medium (the adsorbate) to its surface.

- Natural adsorbents

A natural adsorbent is any material that is not produced synthetically and has adsorptive properties for chemical species of inorganic and organic origin. They have the advantage of being easily acquired, low cost and some are by-products of industrial processes (Araujo, 2009).

According to Dotto, Vieira, Gongalves, and Pinto (2011), the use of low-cost adsorbents in adsorption is now recognised as an effective and economical method for water decontamination. This has led to the development of many studies using the adsorption process to remove colourants from aqueous solutions using low-cost adsorbents.

In general, lignocellulosic adsorbents are agro-industrial by-products, such as corn husks, soya hulls, sugarcane waste, peanuts and coconut. These agro-industrial by-products are basically made up of cellulose, hemicellulose and lignin (Araujo, 2009).

According to Dobrovol'skii (2006), the lignin found in agro-industrial material has the ability to remove heavy metal ions.

According to Pagnanelli, Mainelli, Veglio and Toro (2003), lignin and cellulose are the main compounds in agro-industrial by-products in which the functional group present in these macromolecules has the ability to adsorb metal ions via ion exchange or complexation.

2.2.5.2 Kinetic models of adsorption

The pseudo-first order kinetic model assumes that the rate of change in the concentration of the adsorbed solute with time is directly proportional to the difference between the amount of solute adsorbed at the equiKbrium, q_e, and the amount adsorbed at any given time t, q_t. The pseudo-second order model assumes that the rate of adsorption depends on the amount adsorbed squared. One of the features of this model is that the adsorbate occupies two active sites of the adsorbent.

The equations developed for the two models are presented below.

Pseudo-first order kinetic model:

$$\frac{dq_t}{dt} = k_f * (q_e - q_t) \tag{2.5}$$

By integrating this equation we obtain,

$$\ln(q_e - q_t) = \ln q_e - k_f * t \tag{2.6}$$

Equation (2.6) can be rearranged into the following non-linear form,

$$q_t = q_e * \left[1 - \exp(-k_f * t)\right] \tag{2.7}$$

Where k_f (min-1) is the pseudo-first order adsorption rate constant, q_e (mg/g) is the amount of solute adsorbed in the equiKbrium and q_t (mg/g) is the amount of solute adsorbed at time t (min). The above equation is known as the Lagergren equation. The graph of $\ln(q_e - q_t)$ versus t leads to a straight line with slope $-k_f$ and ordinate at the origin $\ln q_e$. It is possible to assess the fit of this equation to the experimental data using the correlation coefficient R^2. The closer R^2 is to 1.0, the more likely it is that the adsorption kinetics are pseudo-first order.

Pseudo-second order model:

$$\frac{dq_t}{dt} = K_s(q_e - q_t)^2 \qquad\qquad (2.8)$$

where Ks (g/mg*min) is the pseudo-second order kinetic constant. This equation can be integrated with the initial condition $q_{t=0}$ for t=0 to obtain,

$$q_t = K_s q_e^2\, t/(1 + q_e K_s t) \qquad\qquad (2.9)$$

Linearising equation (2.9) results in the following,

$$\frac{t}{q_t} = \frac{1}{(K_s q_e^2)} + \frac{t}{q_e} \qquad\qquad (2.10)$$

The graph of t/qt as a function of t leads to a straight line with slope $1/q_e$ and ordinate at the origin $1/(K_{sqe})^2$.

2.2.5.3 Adsorption isotherms

The adsorption isotherm is a correlation between the amount of solute adsorbed per unit mass of adsorbent (mg/g) and the solute concentration in the fluid phase (mg/L) when the adsorption equilibrium is reached at constant temperature (Perreira, 2008). In single-component adsorption, only one compound is adsorbed onto the surface of the solid.

If there are two or more components that can occupy the same adsorbent surface, the adsorption is multicomponent, which makes the isotherms more complex. The interactions that occur between the different adsorbate species in the fluid phase in this case are of fundamental importance (Schwanke, 2003).

In an adsorption process, temperature mainly affects the kinetics and equilibrium of adsorption. An increase in temperature causes an increase in the adsorption rate constant and the rate of external and internal diffusion of the adsorbate molecules due to a decrease in the viscosity of the solution. Temperature variation also affects the adsorption equilibrium constant for a given adsorbate. Equation (2.11) based on a material balance to the solute is used to calculate the amount adsorbed by the solid at equilibrium,

$$q_e = \frac{(C_o - C_e) \times V}{m} \qquad\qquad (2.11)$$

In which:

q_e (mg/g) is the concentration of the solute in the solid phase after equilibrium has been reached.

C0 (mg/L) is the initial concentration of the solute in the fluid phase.

C_e (mg/L) is the solute concentration in the fluid phase after equilibrium has been reached.

V (L) is the volume of the fluid phase in contact with the adsorbent. m (g) is the mass of the adsorbent.

- Types of isotherms

Adsorption isotherms are classified according to the shapes of their curves, as illustrated in Figure 2.1:

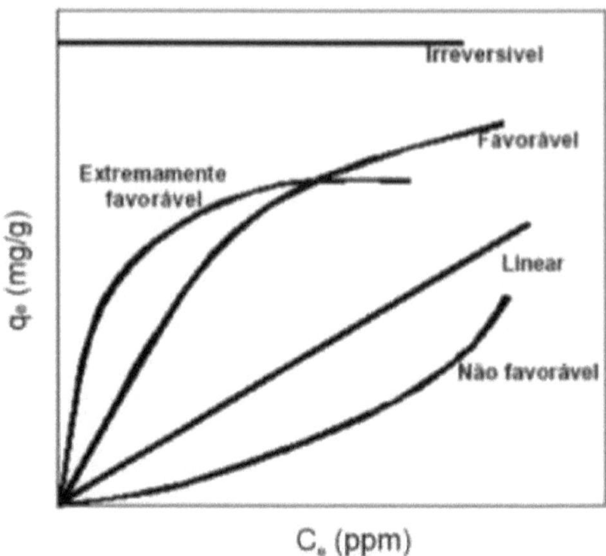

Figure 2.1: Common forms of adsorption isotherms
Source: McCabe, Smith and Harriot (1993)

Some of the most common forms are shown in Figure 2.1. The linear isotherm passes through the origin and the amount adsorbed is proportional to the concentration in the fluid. In favourable isotherms, the slope of the isotherm decreases as the solute concentration in the fluid increases. In unfavourable isotherms, the slope of the isotherm increases as the solute concentration in the fluid increases (Barros, 2005 cited by Melo, 2009).

The simplest isothermal models used in the mathematical modelling of liquid adsorption are the Langmuir monolayer and the Freundlich multilayer (Santos, 2016).

Langmuir isotherm

The simplest representation of the adsorption equilibrium is the Langmuir isotherm, which corresponds to a highly idealised type of monolayer adsorption. According to Santos (2016), the basic considerations of the Langmuir model are:

- On the surface of the solid, there are a finite number of adsorption sites for molecules;
- Each site can only adsorb one molecule;
- There is the same adsorption energy at each location;
- There is no interaction between the adsorbed molecules and their neighbouring sites.

According to these considerations, Langmuir's isotherm can be mathematically deduced from stoichiometric thermodynamics. The general formula for the isotherm applied to the adsorption of liquids is:

$$q_e = \frac{Q_{max} K_L C_e}{1 + K_L C_e} \qquad (2.12)$$

In which:

q_e is the amount of solute adsorbed per gram of adsorbent in the equiKbrium (mg/g);

16

Q_{max} is the maximum adsorption capacity (mg/g);

KL and the adsorbate/adsorbent interaction constant (L/mg);

C_e is the concentration of the solute in the fluid phase at equilibrium (mg/L).

Equation (2.12) is often rearranged into linear forms to determine the values of KL and Q_{max}, as illustrated in the equations below (Nascimento, Lima, Vidal, Melo, & Raulino, 2014):

$$\frac{1}{q_e} = \frac{1}{Q_{max}} + \frac{1}{K_L Q_{max} C_e} \tag{2.13}$$

$$\frac{C_e}{q_e} = \frac{C_e}{Q_{max}} + \frac{1}{K_L Q_{max}} \tag{2.14}$$

$$q_e = Q_{max} - \frac{q_e}{K_L C_e} \tag{2.15}$$

$$\frac{q_e}{C_e} = K_L Q_{max} - K_L q_e \tag{2.16}$$

Freundlich isotherm

The Freundlich isotherm model is based on adsorption on heterogeneous surfaces. The presence of different functional groups and the various adsorbent-adsorbate interactions are caused by the heterogeneity of the surface (Nogueira, 2010).

The Freundlich equation is given by

$$q_e = K_F C_e^{1/n} \tag{2.17}$$

Applying logarithms to equation (2.17) gives us

$$ln q_e = ln K_F + \frac{1}{n} ln C_e \tag{2.18}$$

In which:

q_e is the amount of solute adsorbed per gram of adsorbent in the equiKbrium (mg/g);

C_e is the concentration of the solute in the fluid phase at equilibrium (mg/L);

1/n and a constant related to the intensity of adsorption;

K_F is the Freundlich constant which relates the adsorption capacity and the number of adsorption sites [(mg/g)/(mg/L)$^{A(1/n)}$)].

2.3 Use of Moringa oleifera Lam in sewage treatment

Moringa oleifera Lam is a plant that has been trialled many times in the treatment of raw water and wastewater, so once again it has been trialled in this work for the treatment of domestic sewage mixed with industrial effluent.

According to Cysne (2006), the taxonomic classification of *Moringa oleifera* is as follows:

Division: *Magnoliophyta*

Class: *Magnoliopsida*

Subclass: *Dilleniidae*

Order: *Capparidales*
Family: *Moringaceae*
Genre: *Moringa*
Species: *Moringa oleifera* Lam

2.3.1 Moringa

According to Ndbigengesere and Narasiah (1998) natural coagulants of plant origin were used in water treatment before chemical salts and the use of natural coagulants has been progressively reduced.

Nowadays, there has been a resurgence of interest in these coagulants because they are biodegradable and safe for human health (Okuda, Baes, Nishijima, & Okada, 1999). Other authors such as Ndabigengesere, Narasiah, and Talbot (1995) have examined the properties of cactus in water treatment. Several studies have been carried out to evaluate the activity of compounds present in *Moringa oleifera* Lam seeds.

Moringa oleifera Lam belongs to the Moringaceae family and is a perennial species that originated in northern India. Today, Moringa is cultivated throughout the Middle East as well as along the tropical chain and was introduced to Africa at the beginning of the 20th century (Foidl, Makkar, & Becker, 2001).

According to Correa (1984), the species is widely distributed in the rest of the world, mainly in countries such as Pakistan, Singapore, the Philippines, Thailand, Malaysia, Egypt and Nigeria. *Moringa oleifera* Lam adapts easily to various types of soil, especially well-drained clay, but it cannot withstand prolonged flooding. This plant favours neutral to slightly acidic soil. However, it has been observed to react well when introduced to Pacific atolls, which generally have a pH of over 8.5. It does best at a temperature of between 26 and 40 °C, withstands an annual rainfall of around 500 mm and grows well at an altitude of 1000 m above sea level (Rosa, 1993).

Moringa oleifera Lam is a multi-purpose plant species that utilises all parts of the plant, especially the leaves, green pods, flowers and seeds. The protein content of the leaves can reach 27% and they are rich in vitamin A and C, calcium, iron and phosphorus. In terms of health, *Moringa* leaves are used to help children and adults recover from malnutrition. They are also used externally for inflammation.

Moringa oleifera Lam has medicinal fungi in all its parts. The roots are considered to be stimulants and diuretics, and the flower tea is often used to cure colds. The green pods are eaten as cooked vegetables and the ripe seeds can be roasted to make flour. According to Silva and Matos (2010), the seeds produce an excellent oil that can be used in food. The cosmetics industry uses *Moringa* seeds because of their high oil content.

In other African countries and particularly in Sudan, women traditionally use broken or crushed *Moringa oleifera* Lam seeds to clarify murky water.

In the laboratory and in research projects, the efficiency of this process was demonstrated by placing a certain amount of raw water in a 10 to 20 litre container, adding the pulp of one to three macerated *Moringa* seeds (depending on the quality of the raw water) for each litre of water to be clarified. The mixture was left to stand for two hours, the supernatant removed with a siphon and the precipitate discarded.

Publication of studies on the application of the active extract and seeds (with or without the husk) of *Moringa oleifera* Lam began in the 1980s, in which various studies on the treatment of raw water for human consumption or other uses, wastewater and industrial effluents were reported.

When it comes to treating wastewater or raw water intended for human consumption, it is used in the coagulation-flocculation process. Studies have shown that using either the seed or its aqueous extract to remove turbidity was 80 to 99 per cent efficient (Okuda *et al.*, 1999). According to Ndbigengesere and Narasiah (1998), seed tests carried out with or without shells were highly effective.

In relation to the treatment of industrial water by coagulation-flocculation, Bhatti *et al.* (2007) or Heredia and Martin (2008) mention the beneficial use of *Moringa oleifera* Lam seeds in the removal of metallic compounds such as copper, iron, zinc, aluminium, cadmium, nickel and chromium by the coagulation-flocculation process.

Moringa oleifera Lam seeds have also been used as adsorbents in wastewater treatment.

According to Akhtar *et al.* (2007), *Moringa oleifera* Lam seed used as an adsorbent has the power to remove some organic compounds such as benzene, ethylbenzene, isopropylbenzene and toluene.

In the coagulation-flocculation and adsorption processes, the application of *Moringa oleifera* Lam seeds proved effective in removing different compounds.

2.3.2 Properties of Moringa oleifera Lam

Moringa oleifera Lam is composed of a cationic protein with a molecular weight of 6500 Dalton, which constitutes the active coagulation agent (Gassen, Gassenschmidt, Jany, Tauscher, & Wolf, 1990). The coagulation/flocculation mechanism caused by this protein resembles the mechanism caused by polyelectrolytes, which are organic compounds of natural or synthetic origin, characterised by polymerised chemical compounds that have large carbon chains made up of repeating units, formed of macromolecules weighing between 5,000 and 1,000,000 units. The carbon chains have some points with a deficiency or excess of electric charges, or with positive or negative symmetries (Franco, 2010).

According to Jahn (1988), the coagulant compound of *Moringa oleifera* Lam, after being isolated and identified, showed that it is made up of six polypeptides of various amino acid units. Of the six amino acids detected, the ones present in greatest quantity were: arginine, proline, methionine and glutamic acid.

According to the study and observations of Gueyard *et al.* (2000), Moringa's coagulating action is thought to be related to an amodium compound. Its seeds contain between 8 and 10 per cent glucosinolates, constituting a homogeneous class of combinations of natural thiosaccharides (see Figure 2.2). Natural thiosaccharides can be hydrolysed by myrosinase (thioglucoside glucohydrolase) to produce D-glucose, particularly isothiocionates.

Figure 2.2: Structure of glucosinolate present in Moringa oleifera Lam seed Source: Gueyard *et al.* (2000)

According to Ndibewu, Mnisi, Mokgalaka, and Mccrindle (2011), *Moringa oleifera* Lam's adsorption capacity is highly favoured because most of it contains considerable amounts of cellulose interlinked with lignin in its structure.

It should be noted that lignin is a heterogeneous molecule of a biopolymer complex that has several different functional groups, such as methoxyl, hydroxyl aliphatic, carboxyl and phenolic (Boeriu, Bravo, Gosselink, & Van Dam, 2004).

Lignin has an aromatic, three-dimensional polymer structure with an apparent infinite molecular weight, making biosorption one of the most promising water treatment techniques for removing organic compounds through the use of *Moringa oleifera* Lam as a low-cost adsorbent (Ndibewu *et al.*, 2011).

Finally, other studies have revealed that the cotyledons of *Moringa oleifera* Lam contain an antimicrobial substance that enhances the effect of biological water treatment (Ghebremichael, Gunaratna, Henriksson, & Dalhammar 2005).

As mentioned above, *Moringa oleifera* Lam is a plant with a non-negligible nutritional value, as shown in Table 2.1.

Table 2.1: General characteristics of Moringa by-products used in feed

Content	Pods	Leaves	Leaf dust
Humidity (%)	86,9	75,0	7,5
Calories/100 g	26	92	205
Protein (g/100 g)	2,5	6,7	27,1
Fat (g/100 g)	0,1	1,7	2,3
Carbohydrate (g/100 g)	3,7	13,4	38,2
Fibre (g/100 g)	4,8	0,9	19,2
Mineral (g/100 g)	2,0	2,3	27,1

Source: Price and Davis (2000)

2.3.3 Ecological aspects of Moringa oleifera Lam

The *Moringa oleifera* Lam plant is considered ecologically viable due to its various applications as an alternative to chemical products. When used as a coagulant or adsorbent, this plant has the ability to reduce the risks associated with the accumulation of non-degradable chemical compounds that are harmful to human, animal and environmental health. The plant's properties, including coagulation, flocculation and adsorption, are notable for their ability to clean contaminated water, reducing turbidity, toxicity and microbial load (Gamez, delRisco, & Cano, 2015).

The treatment of raw water or sewage with *Moringa oleifera* Lam seeds ultimately produces a sludge that settles and can be used as a biofertiliser. This represents an additional advantage in rural areas (Mangale, Chonde, & Raut, 2012).

It should also be noted that the use of *Moringa oleifera* Lam seeds in the treatment of coffee fermentation effluents has served as an environmentally viable alternative, since coffee production generates wastewater rich in organic nutrients that are harmful to aquatic ecosystems (Gade, Buchberger, Wendell, & Kupferle, 2017).

In addition to the seeds, the biomass obtained from the husks itself can be used as a promising low-cost compound for treating water and effluents by adsorbing heavy metals from solid agricultural waste (Reddy, Harinath, Seshaiah, & Reddy, 2010a). As *Moringa oleifera* Lam has the capacity to treat aqueous effluents, it has become an alternative for improving public health in socially neglected communities.

2.3.4 Results of previous studies

It was mentioned earlier that *Moringa oleifera* Lam seeds have an active component made up of high molecular weight protems that can act as natural organic polymers. As the solubility of the protems increases with salt concentration for low values of the ionic strength of the salt, various salt extraction methods have been used with NaCl and KCl in order to improve the solubility of the protems.

solubility of the active compound in aqueous solutions (Okuda *et al.*, 2001).

This extraction technique with saline solution provides an increase in extraction efficiency, guaranteeing an increase in coagulation capacity of around 7.4 times compared to extraction with distilled water alone (Okuda *et al.*, 2001). The use of NaCl saline solution is preferred because it does not allow for an increase in Dissolved Organic Carbon (DOC) in the solution, whereas the use of distilled water extraction allows for an increase in DOC in the solution (Okuda *et al.*, 2001).

Many research results prove the effectiveness of *Moringa oleifera* Lam seed extract as a coagulating agent in removing turbidity and coliforms from wastewater and public water supplies.

Chagas *et al.* (2009) went so far as to improve the removal of suspended solids present in recirculation water from the peeling or pulping of coffee fruits and in dairy wastewater, respectively. They concluded that *Moringa oleifera* Lam seed extract showed great potential for application as a natural and alternative coagulant in wastewater treatment.

Silva (2009) carried out studies using natural coagulants in the treatment of effluents from the textile industry and noted that, compared to conventionally used chemical coagulants, *Moringa oleifera* Lam was a promising alternative in the physical-chemical treatment of wastewater, and can be used as an aid in primary treatment, because it is able to provide an increase in the efficiency of decanters by removing suspended solids. Muyibi and Evison (1995) pointed out that many studies have been carried out to separate and characterise the powerful coagulating active ingredient found in *Moringa oleifera* Lam seeds. It was isolated from water and was very convincing in treating water with high turbidity, while for water with low turbidity, it was inefficient. It was observed that the residual turbidity in the treated samples increased as the initial turbidity decreased. This is due to the limited action of *Moringa oleifera* Lam's coagulating agent, especially in water with low turbidity, as well as the need to determine an optimum dosage. It should also be noted that the coagulating activity of *Moringa oleifera Lam* is more appropriate for waters with high turbidity (Okuda *et al.*, 2001).

According to Jahn (1988), various studies have shown that there is no toxicity to humans or animals and that *Moringa oleifera* Lam seeds have coagulating and bactericidal properties.

There is a wide range of applications for *Moringa oleifera* Lam, depending on the area in which it is used. Some have an environmental connotation, as in the case of its coagulation properties, adsorption of metal ions, and the bactericidal action of its seeds in water treatment (Monaco *et al.*, 2010). The team of Amaral, Rossi, Barros,

Lorenzon, and Nunes (2006) also described the role of *Moringa oleifera* Lam seeds as a coagulant and bactericide in water treatment without solar radiation, showing no efficiency in removing pathogenic organisms such as *Escherichia coli*.

Gupta, Dubey, Kannan, and Flora (2007) studied the treatment of water contaminated by metal ions and described the use of the adsorption properties of *Moringa oleifera* Lam seeds in the removal of arsenic from contaminated water, which contaminated people by ingesting groundwater in West Bengal, India.

In the coagulation-flocculation or adsorption processes using the seed, the removal efficiency of the different compounds varies depending on the compound and its content in the sample. Considering, for example,   the results obtained in the study by Akhtar *et al.*   (2007),   the removal of isopropylbenzene was considered the most efficient, demonstrating the high potential for industrial application in low-cost wastewater treatment.

According to Mendes and Coelho (2007), the results obtained using *Moringa oleifera* Lam seeds to remove silver and manganese were particularly effective in waters with a high content of these metals.

Many other studies and investigations have been carried out using *Moringa oleifera* Lam seeds as an adsorbent for removing metals from aqueous effluents, such as:

- Sharma *et al.* (2007) carried out a comparative study of the biosorption of Cd (II), Cr (III) and Ni (II) using crushed *Moringa* seeds as in the case of arsenic removal and proved that the use of *Moringa oleifera* Lam seeds as a low-cost biosorbent is an environmentally sound method for removing cadmium from aqueous media;

- Kumari, Sharma, Srivastava, and Srivastava (2005) revealed the adsorption properties of crushed *Moringa oleifera* Lam seed powder in the removal of As (V) from aqueous solutions;

- Mataka, Henry, Masamba, and Sajidu (2006) presented the first reports on the use of *Moringa stenopetala* of African origin to reduce toxicity and on the preliminary interaction of Pb (II) ion with the functional groups present in *Moringa oleifera* Lam and *Moringa stenopetala*.

According to Akhtar *et al.* (2007), *Moringa oleifera* Lam pods used as a biosorbent have the capacity to remove organic compounds (benzene, toluene, ethylbenzene and cumene), with maximum adsorption capacities of 625 mg/g for benzene, 830 mg/g for toluene, 1061 mg/g for ethylbenzene and 1081 mg/g for cumene being reported, with removal efficiencies ranging from 65 to 85%.

Reddy *et al.* (2010b), using *Moringa oleifera* Lam bark in the biosorption of Pb from aqueous solutions, obtained the result that the bark had an adsorption capacity of 34.6 mg/g related to around 99% removal of Pb.

The use of *Moringa oleifera* Lam bark as a biosorbent for removing Ni (II) from aqueous solutions was successful and it is suitable for use in the separation of Ni (II) from aqueous solutions (Reddy, Ramana, Seshaiah, & Reddy, 2011).

3 METHODOLOGY

The ZEE WWTP was designed to biologically treat domestic sewage mixed with industrial effluent following a process with a preliminary anaerobic treatment stage in upflow sludge blanket reactors (in English terminology known as UASB reactors), followed by an aerobic treatment stage using a Submerged Aerated Biological Filter (SABF) with nitrification capacity, including an anoxic chamber for denitrification. But it turns out that UASB reactors have cracks since they were built, so they don't work. So the effluent coming out of the sandboxes goes straight into the FBAS.

In this work, *Moringa oleifera* Lam was used on the one hand for coagulation-flocculation to remove turbidity and on the other hand as an adsorbent to adsorb organic materials and heavy metals. For coagulation-flocculation, ground raw seed with and without shells and active extracts (aqueous and saline) of seed without shells were used. Seed pods and husks were used for adsorption.

The previous chapter mentioned various laboratory methods for treating the seed and extracting its active compound for use as a coagulant and for treating the pods and husks for use as an adsorbent. To be used as a coagulant in this work, the seed was treated following the procedure described by Price and Davis (2000), its aqueous and saline extracts were obtained following the procedures described respectively by Katayon *et al.* (2006) and Okuda *et al.* (1999). As adsorbents, pods and husks were treated following the mercerisation processes described by Gurgel *et al.* (2007) and Gupta et *al.* (2013).

3.1 Waste water to be treated

In this work, the wastewater samples were taken at the ZEE treatment plant, built in the industrial centre of Viana.

The sewage treatment plant receives industrial effluent that has undergone prior treatment at the production site to prevent the pollution of nature by the discharge of industrial fluids. It is mixed with domestic sewage as it flows through the collection network.

The industrial centre of Viana produces effluents of a different nature and quality depending on the domestic waste resulting from human activity (toilets, offices and restaurants), industrial waste of a predominantly organic nature and industrial waste of a chemical and mineral nature.

By nature, the effluent from an Industrial Centre has variable qualitative characteristics and requires that the treatment process used offers the possibility of adjustment, when necessary, through intervention in the operation.

According to Metcalf and Eddy (1991), sewage is classified as strong, medium and weak, as shown in Table 3.1.

Table 3.1: Physico-chemical characteristics of sewage

Features	Weak	Medium	Strong
BOD5 (mg/L)	110	220	400
COD (mg/L)	250	500	1000
TOC(mg/L)	80	160	290
Total nitrogen (mg/L)	20	40	85
Organic nitrogen (mg/L)	8	15	35
Total Phosphorus (mg/L)	4	8	15

Org Phosphorus (mg/L)	1	3	05
PhosphorusInorg (mg/L)	3	5	10
Chlorides (mg/L)	30	50	100
Sulphates (mg/L)	20	30	50
Oleose Fats (mg/L)	50	100	150

Source: Metcalf and Eddy (1991)

Effluent from the Viana industrial centre can be discharged into the collecting network when it meets the conditions shown in Table 3.2 (ECOSAN, 2012):

Table 3.2: Criteria and standards for discharge of liquid effluents into the WWTP collector (ZEE)

Parameter	Acceptable value
pH	6,0 e 9,0
Temperature (°C)	<40
Sedimentable materials (mL/L)	20
Oils and fats (minerals) (mg/L)	20
Vegetable oils and animal fats) (mg/L)	50
Solvents, petrol, light oils and explosive or flammable substances	Not Detectable
Substances causing blockage of pipework or interfering with operation	Not Detectable
Substances toxic to biological processes	Not Detectable
Total arsenic (mg/L)	0,5
Lead (mg/L)	0,5
Cadmium (mg/L)	0,2
Dissolved copper (mg/L)	1,0
Hexavalent chromium (mg/L)	0,1
Total mercury (mg/L)	0,01
Total silver (mg/L)	0,1
Total selenium (mg/L)	0,3
Total chromium (mg/L)	0,5
Zinc (mg/L)	0,5
Total tin (mg/L)	0,4
Nickel (mg/L)	2,0
Free cyanide (mg/L)	0,2
Total phenols (mg/L)	0,5
Dissolved iron, Fe^{2+} (mg/L)	15,0
Total fluoride (mg/L)	10,0
Sulphide (mg/L)	1,0

Source: ECOSAN (2012)

3.1.1 Treatment process at the EEZ WWTP

According to ECOSAN (2012), the ZEE WWTP has adopted a treatment process based on prior anaerobic treatment in a UASB (Upflow Anaerobic Sludge Blanket) reactor, commonly known as RAFA (upflow anaerobic reactor), followed by an aerobic treatment stage in a Submerged Aerated Biological Filter, with nitrification capacity, and an anoxic chamber for denitrification.

Before biological treatment, preliminary treatment is carried out by harrowing and desanding. The excess sludge from the aerobic process is extracted from the decanters and recirculated to the UASB reactor for anaerobic digestion, together with the sludge synthesised in the anaerobic process. The digested sludge is discarded from the UASB and dewatered in a drying bed before disposal.

The aerobic treatment takes place in a segmented tank with two chambers: the first segment houses the anoxic chamber for denitrification and the second chamber is occupied by the Submerged Biological Aerated Filter (SBAF), where the biomass settles on support elements such as plastic rings. These rings are kept suspended inside the chamber by compressed air supplied from the blowers.

After treatment in the FBAS, the treatment stage follows in the secondary decanter, with the effluent from this decanter being discharged upstream into the Mulenvo River. Figure 3.1 shows a simplified diagram illustrating the sequence of sewage treatment stages at the ZEE WWTP. A detailed flowchart of the treatment process is attached.

Figure 3.1: Sewage treatment scheme at the EEZ WWTP

It should be noted that all effluent generated by any industry in the area, before entering the EEZ's wastewater collection network, undergoes pre-treatment according to its nature, at the place where it is generated, in order to fulfil the languishing criteria shown in Table 3.2.

Industries that generate chemical and mineral effluents, for example, have treatment facilities located on their own land, using the processes of neutralisation, coagulation, flocculation, precipitation, oxidation-reduction, decantation and filtration alone to make the effluent acceptable for the sewage system. On the other hand, industries producing effluents of an organic nature require the removal of substances that inhibit biological treatment.

Some of the treatment stages are described in more detail below:

3.1.1.1 Equalisation

The main task of equalisation is to regulate the flow rate, because the water treatment process needs a constant flow rate. If the flow rate varies too much throughout the process, this is harmful because it makes it impossible for the decanter's pH correction tank to function normally. Equalisation tanks homogenise an effluent to make it uniform. There are several reasons why they should be implemented, such as: minimising operational problems caused by variations in the characteristics of the effluent and shocks caused by overloading the system, improving biological treatment, diluting inhibiting substances, stabilising the pH and improving the final quality of the effluent to be treated (Silveira, 2010).

The effluent retention time in the equalisation tanks is around 19 hours based on the measured flow (ECOSAN, 2012).

3.1.1.2 Preliminary treatment

Preliminary treatment consists of removing solids from the wastewater by mechanical or physical processes. These solids can be removed manually or mechanically from the liquid flow in order to avoid blockages or clogging in the units of the following operations. Sand and earth, which are the predominant inorganic solids, are removed in desanders or sandboxes.

A. Mechanised fine harrowing

The wastewater treatment plant has fine grids to remove floating or dispersed materials that could cause damage or blockage to the process units, namely the anaerobic reactor and the flow distribution lines. At the ZEE wastewater treatment plant, the channels where the grates are installed have manually operated sliding inlet and outlet gates that allow the flow to be blocked when any of the grates is taken out of operation (ECOSAN, 2012).

The material trapped in the grids is removed by cleaning devices and then unloaded onto a conveyor belt that takes it to the rubbish bins.

The grids used are thin mechanised grids made up of parallel metal bars with 3 mm spacing. They are installed upstream of the flow meter (Parshall channel).

According to Silveira (2010), Parshall channels are used for flow measurement. Through their throttled and raised part, they establish a relationship between the flow rate and the height of the water lamina in that region for a given vertical section upstream. Channels reduce head loss and are necessary for reading flows, as shown in Figure 3.2.

Figure 3.2: Parshall channel

B. Desanding

The purpose of desanding is to remove the sandy granular material from the sewage that has escaped removal through the fine grids.

The desander is made up of two parallel channels that allow the particles inside to settle during their journey. The two channels work independently, so that when one is in operation, the other is being maintained and cleaned.

In fact, the combination of these two operations (grading and desanding) adequately preconditions the sewage for subsequent treatment stages, making it possible to avoid or minimise problems of pipe abrasion and excessive equipment wear. Figure 3.3 shows the screw-type bottom scraper and sand lift.

Figure 3.3: Archimedes screw

Each de-sanding unit consists of sandboxes and a sand removal system made up of a bottom scraper and a screw-type sand lift.

The sandboxes are fed by a common channel which receives the sewage from the fine harrowing and which, due to geometric symmetry, provides equal distribution of flow between the units. Each box will be emptied when necessary using portable submersible pumps.

According to ECOSAN (2012), the desanded effluent, through the effluent channel common to the two sandboxes, is sent to a distribution box, where it is discharged freely into proportionally segmented spillways so that 25 % of the desanded sewage flow is equally distributed between the anoxic chambers that are part of the aerobic biological treatment.

Likewise, the remaining flow (75 per cent) is also distributed, via this distribution box, between the two intermediate treatment units of the upflow anaerobic sludge blanket (UASB) type.

The supply to each anaerobic reactor and anoxic chamber can be interrupted by closing one of the stop-log gates located downstream of each spillway.

The outlet gate for the by-pass is installed in the WWTP inlet cell and the diverted flow will be conveyed to the final discharge structure via a system of gravity pipes.

This distribution box is also used to return excess sludge to be stabilised in the UASB reactors.

3.1.1.3 Intermediate treatment

According to ECOSAN (2012), intermediate treatment, which receives 75 per cent of the flow of raw sewage that has been desanded, is carried out by RAFA or UASB sludge blanket ascending flow anaerobic reactors.

The effluent from the desanded sewage distribution box, via 300 mm pipes, reaches the top of the UASB reactors and flows into distribution boxes. The downpipes run from these to a point near the bottom of the tank and ensure that the flow is evenly distributed over the entire bottom of the reactor.

Pouring chutes located at the top of the sedimentation chambers of the UASB reactors, situated just below the surface water level, collect the UASB effluent to be conveyed to the FBAS-type biological reactors as shown in Figure 3.4.

Figure 3.4: UASB or RAFA reactor

The gas produced in the reactors is collected and sent to two flare burners with a unit capacity of 100 Nm^3 /h, located in a fenced area close to the UASB reactors.

The flow rate of the gas produced during digestion is measured in the gas outlet pipe of each reactor.

The sludge collected in the reactors is drained by gravity and collected in underground chutes that discharge into the primary sludge lifts. Each lift is made up of two constant-speed vertical axis progressive cavity pumps, one of which is operational and the other is a backup pump.

The sludge produced by the treatment (sludge generated in the anaerobic reactor plus surplus sludge from the aerobic treatment) is stabilised and removed from the UASB with a solids concentration of around 3.5% and pumped to the drying beds.

3.1.1.4 Secondary treatment

The flow to secondary treatment is made up of two parts:

- 25% of the sewage flow diverted directly into the aerobic treatment anoxic chambers to provide the necessary supply of easily biodegradable BOD for the denitrification process to take place under suitable conditions and in an economical manner;

- 75 per cent of the sewage flow after it has undergone anaerobic treatment.

Secondary treatment involves the following equipment:

• Aerobic reactors

Aerobic biological treatment at the EEZ WWTP is carried out in aerated reactors using the Submerged Aerated Biological Filter (SABF). The FBAS is a reactor in which the activated sludge biomass forms a film on the surface of a support medium. This support medium is made up of small plastic rings, lighter than water, which keep moving and are dispersed inside the reactor. The FBAS is shown in Figure 3.5.

a)   b) c)

Figure 3.5: a) Dumping plastic rings in the FBAS; b) FBAS with rings; c) FBAS in operation

• Secondary decanters

Secondary decantation is applied after biological treatment in the FBAS, with the aim of clarifying the wastewater by sedimenting the biomass. Part of this biomass is then recirculated to the aerobic treatment.

Secondary decanters are cylindrical in shape. The inflow enters through the centre of the tank and the effluent is collected by a flume located inside the tank, which extends along its entire perimeter. This flume is equipped along its entire length with a scum deflector plate and a spillway made up of plates with triangular openings as shown in Figure 3.6.

Figure 3.6: Secondary decanter

The decanter is equipped with a mechanical, centrally-operated sludge scraper , which guides the settled sludge into a recess located near the centre of the tank. The same equipment is fitted with a skimmer to remove the surface sludge and convey it to a sludge pond.

- Sludge Return Elevators

The activated sludge or settled biomass is removed from the bottom of the secondary decanters to be recirculated to the aerobic treatment and discharged into the respective anoxic chamber. This is done via sludge return lifts.

According to Leitao (2018) after secondary treatment, 90 to 98 per cent of suspended solid matter (TSS), 85 to 95 per cent of BOD_5, 30 to 40 per cent of Total Nitrogen and 30 to 45 per cent of Total Phosphorus are removed.

3.1.2 Sample collection

At the ZEE WWTP, samples were taken inside the sandbox, where the liquid effluent (domestic sewage mixed with industrial effluent) after passing through the grading units (coarse and fine) and undergoing desanding and degreasing, as shown in Figure 3.7.

Figure 3.7: Sandpit

It was collected using a bucket tied to a rope that was inserted into the sandbox until it was submerged in the liquid. From the liquid collected in the bucket, 10 five-litre plastic containers were filled after manual stirring to homogenise the liquid (Boaventura, 1997). Collections were made in the morning and afternoon of the same day. Figure 3.8 shows the containers filled with domestic sewage mixed with industrial effluent from the EEZ, which we will refer to as wastewater in this work.

Figure 3.8: Containers with waste water from the EEZ WWTP

The use of plastic containers was due to the quality of the sample collected, which was rich in organic matter (COD over 350 $_{mgO2/L}$) and the quality parameters to be measured in the laboratory.

The collected samples were placed in a cooler with ice and transported to LESRA, where they were kept at 4 °C in a refrigerator. These samples were characterised and then used in the coagulation-flocculation and adsorption experiments. Before use, the samples were removed from the refrigerator, shaken and exposed to the ambient temperature of the laboratory.

3.2 Wastewater characterisation

According to Von Sperling (1996), urban wastewater or domestic sewage contains approximately 99.9% water. The remaining fraction includes organic and inorganic solids, suspended and dissolved, but also microorganisms. Because of this 0.1% fraction, it is necessary to treat sewage before it reaches a watercourse in order to prevent pollution.

Some characteristics of domestic sewage mixed with industrial effluent produced in the EEZ, according to ECOSAN (2012) are: BOD5 500 mg/L, TSS 500 mg/L, SSV 350 mg/L, NTK 50 mg/L and Phosphorus 10 mg/L.

In order to characterise the composite sample collected at the EEZ WWTP, certain parameters were measured in the laboratory, such as pH, conductivity, turbidity, TSS, COD and metals (Cu, Pb, Zn, Cd, Cr and Ni).

Angolan legislation through Presidential Decree No. 261/11 (2011) establishes emission limit values for wastewater discharge, some of which are listed in Table 3.3.

Table 3.3: Emission Limit Values (ELVs) for wastewater discharge

Parameter	Parameter expression	VLE
pH	Sorensen scale	6,0-9,0
COD	mgO2/L	150
SST	mg/L	60
Turvagao	NTU	-
Conductivity	pS/cm, 20 °C	1000
Fe	mg Fe/L	2,0
Lead	mgPb/L	1,0
Cadmium	mgCd/L	2,0
Chrome	mgCr/L	2,0
Copper	mgCu/L	1,0
Nickel	mgNi/L	2,0
Aluminium	mgAl/L	10
Zinc	mgZn/L	5,0

Source: Presidential Decree 261/11 (2011)

3.2.1 pH determination

The method used to determine this parameter is electrometric (using a potentiometer and a combined glass electrode as a sensor).

The pH was determined using the LESRA procedures.

The pH is measured on the Sorensen scale, reflects the hydrogen concentration of a sample and is given as the average of 3 or 4 readings.

A pH of less than 7 indicates that the liquid solution is acidic, with a pH of 7 the solution is neutral and when it is greater than 7 the solution is alkaline.

It should be noted that this measurement was carried out in the laboratory as soon as the sample arrived.

3.2.2 Determining conductivity

The method used at LESRA is electrometric, based on the *Standard Methods for the Examination of Waters and Wastewaters* manual (Eaton *et al.*, 2005).

This determination is made using a conductivity probe fitted with an automatic temperature compensation system so that the readings recorded refer to the reference temperature of 25°C. The conductivity value expressed in pS/cm or mS/cm is the result of the average of three or four readings shown on the device's display.

A high value for the conductivity of the sample implies a high concentration of ions.

3.2.3 Determination of TSS

Solids are made up of suspended and dissolved substances of organic and/or inorganic composition. Suspended solids are considered to be particles with a diameter of more than 1 pm and dissolved solids are considered to be particles with a diameter of less than 10^{-3} pm.

Solids in water can be suspended, colloidal or dissolved depending on their size. According to Von Sperling (1996), suspended solids range in size from 100-103 pm, colloidal solids range in size from 10-3 to 10^{0} pm and dissolved solids range in size from 10^{-6} to 10^{-3} pm.

The gravimetric method was used to determine TSS after filtering the sample. In addition to vigorously shaking the container containing the sample, a peristaltic pump was used to dispense a portion of the sample under magnetic stirring into a beaker.

A 1.2 pm glass fibre filter, previously washed with distilled water and dried in an oven at 103-105° C until constant weight, was used to filter the sample under vacuum. The filter containing the residue after filtration was taken to the oven at Ю3- 105 °C for one hour. After this time, the filter with the residue was removed from the oven, left to cool in the desiccator for an hour and weighed. These drying, cooling and weighing operations were repeated until a constant weight was reached. The increase in the weight of the filter paper represents the total suspended solids.

The result calculated by Equation (3.1) is expressed in mg of suspended solids per litre of sample.

$$Sólidos\ Suspensos\ Totais\ (^{mg}/_{L}) = \frac{(B - A) * 1000}{Volume\ de\ amostra\ (ml)} \qquad (3.1)$$

In this formula, A is the weight of the empty filter paper (mg) and B is the weight of the filter paper with dry residue (mg).

3.2.4 Determining turbidity

Turbidity determination was based on the Nephelometric method.

The sample to be analysed (20 mL), previously shaken, was placed in the measuring cell. This cell was washed three or four times with the sample to be analysed before inserting the 20 mL sample for reading. The cell with the sample was inserted into the previously calibrated turbidimeter to read the turbidity.

The turbidity result was given as the average of three or four readings.

The unit of measurement for turbidity obtained using this technique is NTU (Nephelometric Turbidity Unit).

3.2.5 COD determination

The Chemical Oxygen Demand (COD) is a measure of the amount of oxygen needed to chemically oxidise organic matter. This organic matter, which includes biodegradable material, when discharged into receiving bodies causes a decrease in the concentration of dissolved oxygen in the water, deteriorating water quality or making aquatic life impossible (Giordano, 2020).

COD was determined following the modified procedure of Raposo *et al.* (2008) cited by Leitao *et al.* (2017). The modified closed reflux method is based on the *American Public Health Association-American Water Works Association-Water Pollution Control Federation* (APHA-AWWA-WPCF) reference method (Eaton *et al.*, 2005).

This modified procedure is recommended for solid or aqueous samples with a high TSS content and COD in the range of 40 to 900 mgO2/L.

The sample as it was was disintegrated in the ultrasonic cooker for 15 minutes to break up the particles and thus facilitate the process of homogenising the sample before pipetting the sample for analysis.

The procedure used is summarised below:

- 2.5 mL of the sample was pipetted into the digestion tube, 1.5 mL of the digestion reagent and 3.5 mL of the catalyst solution were slowly added to the mixture while the digestion tube was shaken.

- Each of the digestion tubes was corked and placed on the heating block at 150 °C for two hours. After digestion, the tubes were removed and cooled to room temperature. The contents of each tube were transferred to 50 mL beakers and 3 drops of ferroma indicator were added to titrate with FAS. The change in colour from blue-green to brownish-red indicated the end of the titration and the volume (mL) of FAS consumed in the titration of each sample was noted. The COD content of the sample in mgO2/L was calculated using the formula given by equation (3.2):

$$CQO = \frac{(B - A) * Molaridade\ do\ FAS * 8000}{2,5} \qquad (3.2)$$

where A is the volume of FAS consumed by the sample and B is the volume of FAS consumed by the blank, both expressed in mL.

To calculate the exact molarity of the FAS, 2.5 mL of 0.25 N potassium dichromate was titrated with the prepared FAS.

The molarity of FAS was calculated using the formula given by Equation (3.3):

$$Molaridade = \frac{0,25 * 2,5}{Volume\ de\ FAS\ gasto\ na\ titulação} \qquad (3.3)$$

3.2.6 Determination of heavy metal content

The heavy metal content was determined using the Flame Atomic Absorption Spectrophotometry method.

For soluble metal determination, the samples were previously filtered using a 0.45 pm cellulose membrane filter in a vacuum filtration system and then acidified. For total metal determination (soluble plus particulate) the sample was acidified without any prior filtration.

According to Santos (2012), to release the metal bound to organic substances, the unfiltered or filtered samples are digested with an acid solution in the digestion block used for COD determination, using COD tubes, under controlled conditions.

The procedure was as follows: 100 mL of sample (raw or filtered waste water, supernatant from the Jar Test or its respective filtrates and the solution after the metal adsorption test) were introduced into the tubes containing boiling regulators. Then 0.6 mL of HNO3 and 0.2 mL of HCl were added. The samples were digested for 30 minutes under closed reflux at 140 °C. Once cooled, they were filtered through a 0.45 pm membrane filter and the filtrates were transferred to polyethylene bottles which were kept in the refrigerator at 4 °C until the date of the analyses.

The standards were all prepared by diluting factory solutions (MERCK) of each metal with a concentration of 1000 mg/L with distilled water.

The atomic absorption spectrophotometer directly provides the absorbance reading, which is linearly related to the metal concentration by means of a calibration line obtained by reading the absorbances of standards of known metal concentration. The result is expressed in mg of metal/L.

Table 3.4 shows the operating parameters for the determination of metals by EAA and their detection limits.

Table 3.4: Operational parameters for metal determination by EAA

Metal	Wavelength Л (nm)	Flame type	Optimum reading range (mg/L)	Detection limit (mg/L)
Cd	228,8	Air-acetylene	0,05-2	0,02
Cr	357,9	Air-acetylene	0,2-10	0,02
Ass	324,7	Air-acetylene	0,2-10	0,02
Fe	248,3	Air-acetylene	0,3-10	0,05
Hg	253,7	Steam cold/argon	0,01-10	0,01
Mn	279,5	Air-acetylene	0,1-10	0,01
Ni	232,0	Air-acetylene	0,3-10	0,02
Pb	217,0	Air-acetylene	1-20	0,05
Zn	213,9	Air-acetylene	0,05-2	0,01

Source: Santos (2012)

3.3 Preparation of Moringa oleifera Lam for use as a coagulant/flocculant and adsorbent

The *Moringa oleifera* Lam seeds used in the wastewater treatment processes were harvested in Bairro Fubu, Camama commune, Talatona district in Luanda. In this work, the treatment processes were implemented for two different uses of the material: coagulation-flocculation and adsorption. For the coagulation-flocculation test, the seed (unshelled and shelled) was used, as well as the aqueous and saline active extracts of the unshelled seed, and for the adsorption process, the seed pods and shells were used. The material collected for the tests is illustrated in Figure 3.9.

 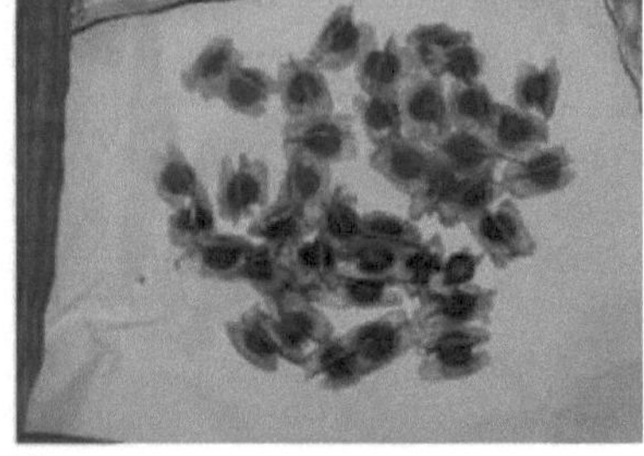

a) b)

Figure 3.9: a) *Moringa oleifera* Lam plant with pods; b) Seeds

Different treatments were given to the parts of the plant used in the different trials.

-   Ground seeds

For use as a natural coagulant, Moringa seeds were treated following the procedure of Price and Davis (2000).

The harvested pods were peeled to remove the seeds, which were dried at room temperature for about 21 days. In this study, both shelled and unshelled seeds were used. The dried shelled and unshelled seeds were ground separately in a coffee grinder as shown in Figure 3.10.

Figure 3.10: Mill used to grind seeds

After crushing, the powder collected from each type of seed (Figure 3.11) was sieved through a 1 mm mesh sieve to obtain the size fraction used in this work (< 1 mm).

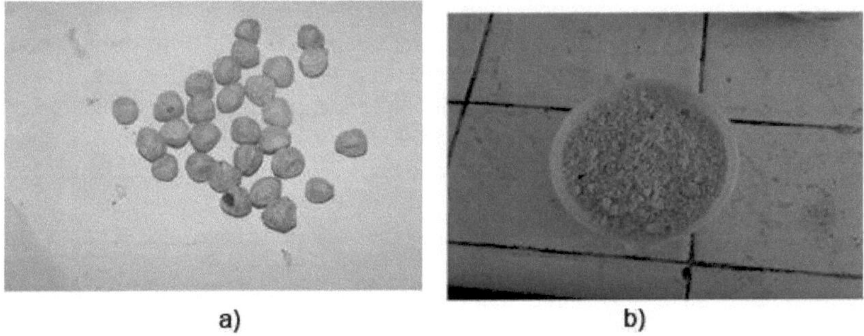

a)   b)

Figure 3.11: a) Seeds without shells (kernels); b) Seed dust

-    Seed extracts

Different extracts were prepared as follows:

- Aqueous extract, prepared according to the procedure described by Katayon and Noor (2006). The seed powder was mixed with distilled water (5 g of powder in 200 mL of distilled water) and stirred for 60 minutes. After this contact time, the dispersion obtained was filtered under vacuum using a 1.2 pm glass fibre filter. The filtrate was placed in a 500 mL flask and topped up with distilled water. The extract obtained under these conditions corresponds to a concentration of 10000 mg of Moringa oleifera Lam/L. Figure 3.12 illustrates the process of obtaining the active compound to be used in the experiments.

a)   b)

Figure 3.12: a) Aqueous mixture of seed powder; b) Vacuum filtration after extraction of the active compound.
compound

- Salt extract: It has been reported by Okuda *et al.* (1999) that the solubility of proteins increases with salt concentration for low values of the ionic strength of the salt.

In this work, the procedure reported by those authors was followed to obtain the saline extract of the active component of *Moringa oleifera* Lam from its seeds.

5 g of the powder obtained from grinding the seeds was added to 100 mL of 1 M NaCl solution. The suspension thus prepared was left to stir for 10 minutes. The solid fraction was then separated by vacuum filtration using a glass fibre filter with a pore size of 1.2 pm. The filtrate was placed in a 500 mL flask and topped up with distilled water. The extract obtained under these conditions corresponds to a concentration of 10000 mg of *Moringa oleifera* Lam/L.

The two types of extract obtained will be used in this work to compare their effectiveness.

As reported by Ribeiro (2010), the aqueous or saline extract of the active component of *Moringa oleifera* Lam is a whitish liquid with an intense odour. There was also a great deal of foaming, which did not make filtering any easier. The extract's shelf life is limited due to the degradation of biodegradable organic matter. For this reason, the extract was produced on the days when the coagulation-flocculation tests were carried out.

- Pods and husks

The pods and husks used as adsorbents in this work were treated following the mercerisation process proposed by Gurgel *et al.* (2007) and Gupta *et al.* (2013). 10 g of adsorbent material (pods and husks) were treated with 250 mL of 30% NaOH solution (300 g NaOH/L). The mixture was stirred for 15 hours, then filtered, washed with deionised water until all the excess sodium hydroxide was removed (until pH 7 was reached) and finally the material was dried in an oven at 60 °C for 24 hours.

Before the mercerisation process, the adsorbents were crushed and sieved through a series of sieves with diameters: 1.6 mm > 1 mm > 0.85 mm. The 1 mm to 0.85 mm size range was chosen for the tests because of its surface area. It is well known that the smaller the size, the greater the surface area, so the smallest possible size was chosen so as not to hinder the sol- liquid separation after adsorption.

3.4 Coagulation-flocculation tests

Following the procedure reported by Ribeiro (2010) used at LESRA, the coagulation-flocculation tests comprise three stages: in the first stage the effect of the coagulant dose at a fixed pH on turbidity is studied, in the second stage the effect of pH is studied using the optimum coagulant dose found in the previous stage and finally in the third stage, carried out at the optimum pH found in the previous stage, the optimum coagulant dose found in the first stage is confirmed. Any of these three stages involves three sequential phases: coagulation, flocculation and sedimentation. Coagulation is carried out with increased agitation in order to promote optimal mixing of the coagulant with the colloidal particles. The added coagulant neutralises the negative surface electric charges of the particles which prevent them from agglomerating. Flocculation is carried out with slower agitation in order to promote the agglomeration of the particles and the formation of flocs. Sedimentation is carried out without agitation so that the flocs formed settle.

The procedure described above was carried out in a device called a "jar test", using 1000 mL of sample in each jar and the seed extract and ground seed as coagulants. The conditions under which the experiments with the different coagulants were carried out are detailed below.

3.4.1 Use of Moringa oleifera Lam seed extract

The procedure was applied to the aqueous and saline extracts of Moringa.

- First stage of the jug test

At this stage, the effect of varying the dose of *Moringa oleifera* Lam aqueous extract on turbidity was studied.

To test the ability to remove turbidity from wastewater using the aqueous extract of *Moringa oleifera* Lam seed, the volume of extract was varied between 4 mL and 28 mL in the different jars of the equipment, corresponding to concentrations of 40 to 280 mg/L of *Moringa oleifera* Lam. The tests were carried out without any pH variation and the natural pH of the wastewater was used in each jug.

- Second stage of the jug test

In this stage, the pH variation between 4 and 12 was studied in order to find the optimum pH value using the optimum dose of *Moringa oleifera* Lam extract found in the first stage.

- Third stage of the jug test

This step consists of the necessary confirmation of the optimum dose of coagulant to be used with the optimum pH value found in step 2. The volume of *Moringa oleifera* Lam extract is varied around the optimum volume found in the first stage, using a narrower range of values.

3.4.2 Use of Moringa oleifera Lam seed powder

The procedure was applied to both unshelled and shelled ground seeds.

- First stage of the jug test

This stage involved studying the effect of varying the dose of *Moringa oleifera* Lam seed on reducing the turbidity of waste water. For the unhulled seed, the dose of seed powder was varied between 40 mg and 200 mg in the different jars of the equipment, corresponding to concentrations between 40 and 200 mg/L. In the case of the seed powder with husk, the dose was varied between 40 and 320 mg, corresponding to concentrations between 40 and 320 mg/L.

In this first stage, the tests were carried out at the natural pH of the wastewater sample.

*   Second stage of the jug test

This stage involves studying the effect of varying the pH on the removal of turbidity from the sample.

The main objective of this stage is to find the optimum pH value, in the range of 4 to 12, at which coagulation-flocculation should be carried out, using the optimum dose of *Moringa oleifera* Lam found in the first stage.

*   Third stage of the jug test

At this stage, the optimum dose of coagulant is confirmed.

With the optimum pH found in the second stage, the dose of *Moringa oleifera* Lam seed powder was varied again, considering a narrower range of values.

For all the coagulants tested, Table 3.5 shows the values of the operating variables used in the sequential coagulation, flocculation and sedimentation phases of each stage of the tests carried out in the jar test rig.

Table 3.5: Operational variables used in the sequential phases of each Jar Test stage

Mixing Quick coagulation speed (rpm)	Mixing Quick coagulation duration (min)	Slow Mix flocculation speed (rpm)	Slow Mix flocculation duration (min)	Time Sedimentation (min)
120	4	30	25	30

Source: Ribeiro (2010)

In each stage of the jar test, a blank test was carried out using the sample at its natural pH without the addition of a coagulant. Also, after the sedimentation phase of each stage, the turbidity, pH and conductivity of the supernatant solution from each jug were measured.

In addition, the quality parameters of the treated wastewater under the optimum conditions for each coagulant were determined.

Figure 3.13 shows the appearance of the wastewater sample mixed with the moringa seed before treatment (a) and after treatment (b).

a)   b)

Figure 3.13: a) Samples before treatment; b) Samples after treatment

3.5 Adsorption tests

The parts of *Moringa oleifera* Lam used as adsorbents in this study were the pod and the seed coat. The adsorption of soluble organic matter and soluble metal in the wastewater sample was investigated. For this purpose, the sample collected from the EEZ wastewater treatment plant was pre-treated at LESRA to pH 12 in order to sediment the particulate matter, as shown in Figure 3.14. The supernatant resulting from this pre-treatment was submitted to the adsorption tests carried out at fixed temperature and pH values. The pH of the sample was adjusted using HCl or NaOH 6 N solutions.

a)   b)

Figure 3.14: a) Raw wastewater from the EEZ WWTP; b) Wastewater after pre-treatment

3.5.1 Determining the zero charge point and the effect of pH on adsorption

The Zero Charge Point (ZCP) is the point at which the adsorbent's surface charge balance equals zero. Each adsorbent has a pH value at which its surface has zero charge. Determining the PCZ is necessary in order to analyse in which pH range the solute is favourably adsorbed (Santos, 2016).

The Zero Charge Point was determined following the procedure of Regalbuto and Robles (2004), in which 50 mg of the adsorbent was mixed with 50 mL of KCl at two different concentrations (0.05 and 0.5 mol/L), with initial pH values between 2 and 10, adjusted with 0.1 mol/L HCl or NaOH solutions. The final pH values were measured after 24 hours under agitation at 150 rpm. The zero load point corresponds to the initial pH value at which the change in pH (final pH - initial pH) is zero.

To assess the influence of the pH of the medium on the adsorption of organic matter by the pods and husks under study, the pH values of the sample were adjusted with 0.1 N HCl and NaOH solutions. pH values of 2, 4, 6, 8 and 10 were tested for both adsorbents. The mass of either adsorbent (pods and husks) used in each test was 50 mg placed in a sachet and immersed in 100 mL of sample contained in a 200 mL conical flask. The tests were carried out under agitation at 150 rpm at 20 °C for one hour. After this time, the sachet with the adsorbent was removed from each flask and the supernatant was analysed to determine the organic matter (COD) content.

3.5.2 Determination of adsorption kinetics

The tests to determine adsorption kinetics consisted of placing 50 mg of adsorbent in contact with 100 mL of the sample in different erlenmeyers placed in a thermostatised shaker with 150 rpm agitation. The flask was removed from the shaker at different times (30 min, 1, 2 and 3 hours). To make it easier to separate the sample from the adsorbent after adsorption, the adsorbent was placed in each erlenmeyer flask in sachets dipped in the sample so as not to touch the bottom of the container. In order to account for the possible desorption of the adsorbate, which is part of the adsorbent's composition, 50 mg of adsorbent was placed in contact with 100 mL of deionised water in different erlenmeyer flasks, removed from the shaker at the same times as the erlenmeyer flasks containing the samples. In order to investigate the possible alteration of the sample during the test time, 100 mL of the sample was placed in an erlenmeyer flask without adsorbent. The tests were carried out at two constant temperatures (20 and 40 °C) and at the pH value of the sample for the most favourable adsorption.

3.5.3 Determining the adsorption equilibrium

The tests to determine the adsorption equilibrium consisted of placing different masses of the adsorbent (50, 100, 150, 200, 300, 500, 1000, 1500, 2000, 2500 and 3000 mg) in contact with 100 mL of the sample in different erlenmeyers placed in a thermostatised shaker with 150 rpm agitation. All the erlenmeyers were removed from the shaker at the end of the equilibration time determined with 70
based on the adsorption kinetics tests. To make it easier to separate the sample from the adsorbent after adsorption, the adsorbent was placed in each flask in sachets dipped in the sample so as not to touch the bottom of the container. In order to account for the possible desorption of the adsorbate, which is part of the adsorbent's composition, the different masses of adsorbent were placed in contact with 100 mL of deionised water in different erlenmeyer flasks, removed from the shaker at the end of the equilibration time. In order to investigate the possible alteration of the sample during the test time, 100 mL of the sample was placed in an erlenmeyer flask without adsorbent. The tests were carried out at two constant temperatures (20 and 40 °C) and at the pH value of the sample for the most favourable adsorption.

4 RESULTS AND DISCUSSION

This chapter presents the main results obtained in the experiments carried out, which were described in the previous chapter, with a brief discussion of the theoretical foundations reported in the literature.

Based on the scientific objectives set for this work, the most relevant results are presented.

4.1 Characterisation of the waste water collected at the EEZ WWTP

After being taken to the laboratory, the wastewater collected at the WWTP was analysed at its natural pH. The values of the parameters measured in four samples are shown in Table 4.1.

Table 4.1: Characterisation of wastewater from the EEZ WWTP

Parameter	Sampling						
	A1	A2	A3	A4	Media	95% confidence interval	VLE
pH	6,58	6,44	6,59	6,51	6,53	6,53 ± 0,10	6,0-9,0
EC (pS/cm)	1917	1923	1911	1932	1921	1921 ± 12	1000
TSS (mg/L)	200	321	167	625	328	328 ± 287	60
Turbidity (NTU)	378	355	354	468	389	389 ± 74	-
COD (mgO2/L)	1412	1811	1868	1718	1702	1702 ±280	150

An analysis of Table 4.1 shows that the average value of conductivity (EC) 1921 pS/cm, TSS 328 mg/L, turbidity 389 NTU and organic matter (COD) 1702 mgO2/L exceed the Emission Limit Value (ELV) established by Angolan legislation (Presidential Decree No. 261/11 of 6 October 2011). Although the average pH value of this wastewater is 6.53, the values of the other parameters, for the most part, do not fall within the range permitted by the aforementioned legislation. Therefore, this wastewater produced in the EEZ cannot be discharged into a watercourse without prior treatment at the WWTP, in order to avoid damaging the environment.

4.2 Coagulation-flocculation tests

The coagulation-flocculation tests were carried out in three stages: stage 1 was used to determine the optimum dose by varying the coagulant dose at the natural pH of the sample; stage 2 determined the optimum pH by varying the pH of the sample and using the optimum dose found in stage 1; and finally stage 3 was used to certify the optimum dose (by varying the dose in a narrower range around the optimum value found in stage 1) with the sample adjusted to the optimum pH found in stage 2. Fresh samples were taken for the tests with each coagulant.

4.2.1 Coagulation-flocculation test with Moringa oleifera Lam seed aqueous extract

The aqueous extract was obtained following the procedure of Katayon and Noor (2006) described in Chapter 3.

4.2.1.1 Determining the optimum conditions for carrying out the tests

The optimum conditions were determined in three stages, as mentioned above. Each stage consisted of coagulation, flocculation and sedimentation phases. At each stage, the pH, conductivity and turbidity values of the sample without coagulant were measured. At the end of each stage, after the sedimentation phase, these parameters were measured in the supernatant of each jug. It should be noted that the same procedure was carried out for all coagulants.

First stage of the jar test: Study of the effect of varying the coagulant dose. This stage was used to determine the optimum dose of coagulant for which the greatest turbidity removal occurred.

Before the jug test, some parameters were measured for this sample and the following results were obtained: natural pH 6.15, conductivity 639 pS/cm and turbidity 357 NTU. The final turbidity value obtained and the turbidity removal percentage as a function of the coagulant dose are shown in Table 4.2 and Figure 4.1.

Table 4.2: Effect of coagulant dose on turbidity removal at the sample's natural pH (step 1)

Dose (mg/L)	Parameter			
	P^H	EC (pS/cm)	Turbidity (NTU)	% redemption
40	6,14	608	216	39,5
60	6,15	610	206	42,3
80	6,11	612	207	42,0
100	6,07	610	198	44,5
120	6,09	610	209	41,5
140	6,06	609	207	42,0
160	6,04	610	203	43,1
180	6,05	585	166	53,5
200	6,20	585	164	54,1
220	5,98	586	163	54,3
240	6,01	585	169	52,7
260	5,98	585	167	53,2
280	5,97	585	170	52,4

Figure 4.1 illustrates the percentage of turbidity removal for different doses of coagulant.

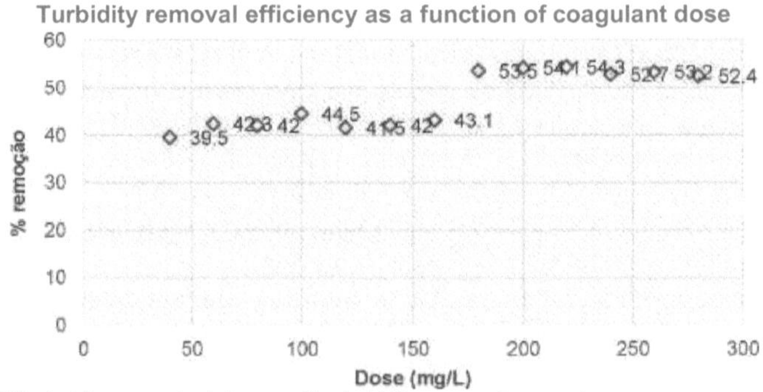

Figure 4.1: Effect of the coagulant dose on the turbidity removal percentage

It was found that the use of the aqueous extract reduced turbidity from 357 NTU to 163 NTU, corresponding to 54.3% removal at a dose of 220 mg/L of coagulant. The initial pH of the sample in each jug was constant and equal to the natural pH of the sample. At the end of the stage, the pH in all the jugs was around 6.1 and the conductivity around 598 pS/cm.

The results in Table 4.2 and Figure 4.1 show that for doses lower or higher than 220 mg/L the final turbidity is higher than 163 NTU and the % turbidity removal is lower at 54.3%. The optimum dose found at this stage was therefore 220 mg/L where the greatest turbidity removal occurred.

- **Second stage of the jug test**: Study of the effect of varying the pH. The aim of this second stage was to find the pH value at which coagulation should be carried out for the greatest removal of turbidity. This pH was determined using the optimum dose of coagulant found in stage 1 (220 mg/L). The initial pH value of each jug was varied between 4 and 12 to cover the alkaline and acidic pH ranges. The sample submitted for treatment had a natural pH of 6.15, conductivity of 639 pS/cm and turbidity of 357 NTU.

The final turbidity values and the turbidity removal percentage as a function of the different initial pH values are shown in Table 4.3 and Figure 4.2.

Table 4.3: Effect of initial pH on turbidity removal with a coagulant dose of 220 mg/L (step 2)

Initial pH	Final pH	EC (pS/cm)	Turbidity (NTU)	% redemption
4	4,07	479	149	58,3
6	6,65	601	169	52,7
8	7,87	745	141	60,5
10	9,77	744	103	71,1
12	11,76	2530	9,45	97,4

Figure 4.2 illustrates the percentage of turbidity removal for the different initial pH values of the sample.

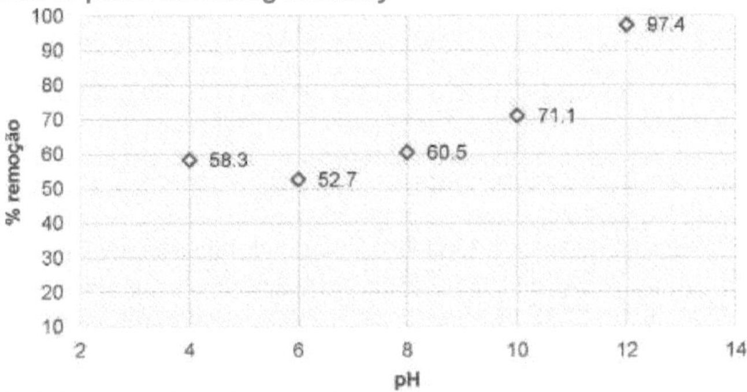

Figure 4.2: Effect of pH on the turbidity removal percentage

Table 4.3 and Figure 4.2 show that turbidity was reduced from 357 NTU to 9.45 NTU, corresponding to 97.4% turbidity removal, using a pH of 12, with a dose of 220 mg/L of coagulant.

At the end of stage 2, the final pH in all the jars was around the initial pH. The final conductivity varied more when compared to its initial value; for acidic pH values there was a reduction in conductivity and for basic pH values there was an increase, this increase being considerable at pH 12.

43

Table 4.3 shows that turbidity had its maximum value (169 NTU) at an initial pH of 6, after which turbidity decreased for increasing pH values.

Third stage of the jar test: Confirmation of the optimum dose of coagulant.

This stage served to confirm the optimum dose of coagulant in tests carried out at the optimum pH (12) found in stage 2. The sample subjected to treatment had an initial turbidity of 357 NTU, initial conductivity of 639 pS/cm and an initial (natural) pH of 6.15.

At the end of this stage, parameters were measured in the supernatant of the treated sample, and the values obtained are shown in Table 4.4 and Figure 4.3.

Table 4.4: Effect of coagulant dose on turbidity removal at an initial pH of 12.

Dose (mg)	Final pH	EC (pS/cm)	Turbidity (NTU)	% redemption
180	11,73	4030	22,9	94
200	11,76	4040	21,8	94
220	11,75	4020	26,0	92
240	11,75	4000	19,7	95
260	11,75	4000	17,8	95

Figure 4.3 illustrates the turbidity removal efficiency as the coagulant dose varies.

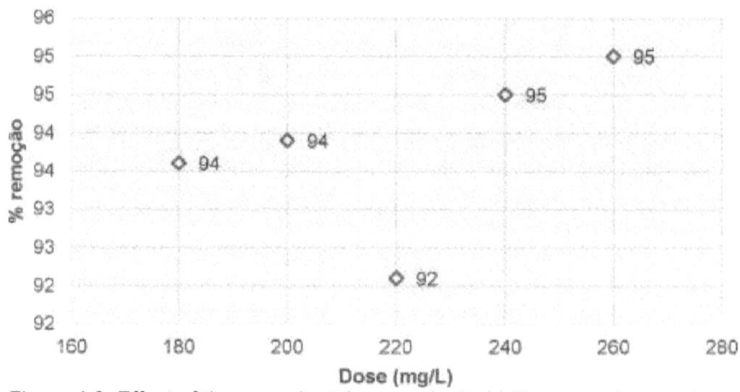

Figure 4.3: Effect of the coagulant dose on the turbidity removal percentage

The results of this last stage of the jug test show that the coagulant reduced turbidity from 357 NTU to 17.8 NTU, corresponding to 95% turbidity removal, with a dose of 260 mg/L at an initial pH of 12. It was also found that conductivity varied considerably from 639 pS/cm to 4000 pS/cm, a greater increase than that recorded in stage 2 with an initial pH of 12 and a coagulant dose of 220 mg/L. The increase in conductivity can be explained by the addition of the pH adjustment reagent and the fact that the coagulant is a cationic protein.

It should be noted that the final optimum dose (260 mg/L) of the *Moringa oleifera* Lam aqueous extract coagulant found in this last stage of the jar test removed 95 per cent of turbidity, and this turbidity removal was lower than that found in stage 2 (97.4 per cent) for the 220 mg/L dose.

The results obtained in the 3 stages of the jar test show that pH is a crucial parameter for achieving good turbidity removal results.

4.2.1.2 Wastewater characterisation after coagulation-flocculation test under optimum conditions

The raw wastewater was subjected to a coagulation-flocculation test using the optimum pH values (pH=12) and coagulant dose (260 mg/L) found in section 4.2.1.1. Table 4.5 shows the values of the parameters measured in the untreated wastewater and in the wastewater treated by the coagulation-flocculation process carried out under the optimum conditions.

Table 4.5: Characterisation of raw wastewater and wastewater treated with *Moringa oleifera* Lam aqueous extract under optimum conditions.

Parameter	Raw wastewater	Treated wastewater
pH	6,15	11,75
Conductivity (pS/cm)	639	4000
Turbidity (NTU)	357	17,8
COD (mgO2/L)	1296	920
TSS (mg/L)	328	-
Pb (mg/L)	< LD	< LD
Zn (mg/L)	0,0286	0,0269
Cu (mg/L)	< LD	< LD
Cr (mg/L)	0,0207	0,0223
Fe (mg/L)	< LD	< LD
Ni (mg/L)	< LD	< LD

The analysis in Table 4.5 shows that the coagulation-flocculation treatment contributed to a 95.0% reduction in turbidity and a 29% reduction in COD. However, there was a considerable increase in conductivity which can be attributed to the nature of the coagulant, which is a cationic protein, and to the reagent used to adjust the initial pH. The treatment had no significant effect on the levels of heavy metals measured. These levels are below the emission limit values and were measured by atomic absorption spectrophotometry on acidified and digested samples as described in Chapter 3. The reduction in COD with treatment can be attributed to the reduction of suspended solids containing organic matter.

4.2.2 Coagulation-flocculation test with saline extract of moringa seed

The saline extract used was obtained following the procedure described by Okuda *et al.* (1999) as mentioned in Chapter 3.

4.2.2.1 Determining the optimum conditions for carrying out the tests

As mentioned above, the procedure was the same for all the coagulants. The optimum conditions were determined in three stages. Each stage consisted of coagulation, flocculation and sedimentation phases. At each stage, the pH, conductivity and turbidity values of the sample without coagulant were measured. At the end of each stage, after the sedimentation phase, these parameters were measured in the supernatant of each jug.

First stage of the jar test: Study of the effect of varying the coagulant dose. This stage was used to determine the optimum dose of coagulant for which the greatest turbidity removal occurred.

As mentioned above, before the jug test, some parameters were measured for this sample and the following results were obtained: natural pH 5.58, conductivity 617 pS/cm and turbidity 345 NTU. The final turbidity value obtained and the turbidity removal percentage as a function of the coagulant dose are shown in Table 4.6 and Figure 4.4.

Table 4.6: Effect of coagulant dose on turbidity removal at the sample's natural pH (step 1)

Dose (mg/L)	2H,	EC (pS/cm)	Turbidity (NTU)	% redemption
40	5,91	976	218	36,8
60	5,89	1183	221	35,9
80	5,87	1371	218	36,8
100	5,94	1574	209	39,4
120	5,95	1786	184	46,7
140	5,95	1928	164	52,5
160	5,94	1927	162	53,0
180	5,98	1942	106	69,3
200	5,9	1943	103	70,1
220	5,87	2490	52,8	84,7
240	5,93	3480	41,5	88,0
260	5,92	3550	48,7	85,9
280	5,9	3650	59,7	82,7

Figure 4.4 shows the turbidity removal efficiency for different doses of the coagulant.

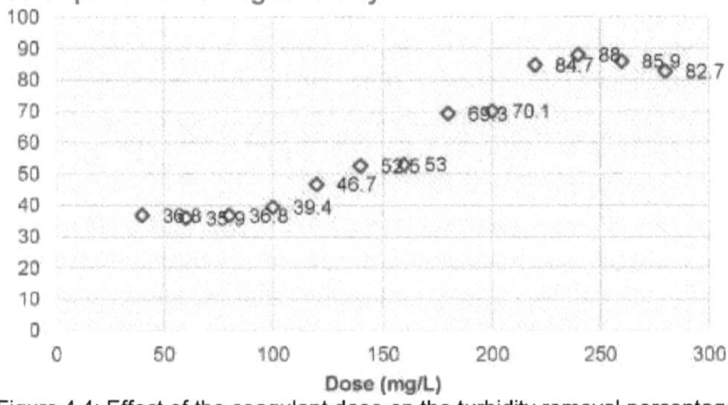

Figure 4.4: Effect of the coagulant dose on the turbidity removal percentage

Table 4.6 shows that the use of the coagulant reduced turbidity from 345 NTU to 41.5 NTU, corresponding to 88% removal at a dose of 240 mg/L. It was also found that at the end of the stage, in each sample jug, the final pH value was around the value of the initial pH 6 and the final conductivity value was higher than the initial conductivity, increasing sharply with the dose of coagulant due to the fact that it is a saline extract from the seed.

The optimum dose found in stage 1 was 240 mg/L, which corresponds to the highest turbidity removal value (88%).

Second stage of the jug test: Study of the effect of varying the pH. The aim of this second stage was to find the pH value at which coagulation should be carried out for the greatest removal of turbidity. This pH was determined using the optimum dose of coagulant found in stage 1 (240 mg/L). The initial pH value of each jug was varied between 4 and 12 in order to cover the alkaline and acidic pH ranges. The sample submitted for treatment had a natural pH of 5.58, conductivity of 617 pS/cm and turbidity of 345 NTU.

The final turbidity values and the turbidity removal percentage as a function of the different initial pH values are shown in Table 4.7 and Figure 4.5.

Table 4.7: Effect of initial pH on turbidity removal with a coagulant dose of 240 mg/L (step 2)

Initial pH	Final pH	EC (pS/cm)	Turbidity (NTU)	% redemption
4	4,18	3880	204	40,8
6	6,03	3450	102	70,4
8	8,94	3510	37	89,2
10	10,14	4290	15,4	95,5
12	12,05	5870	5,51	98,4

Figure 4.5 shows the turbidity removal efficiency for the different initial pH values of the sample.

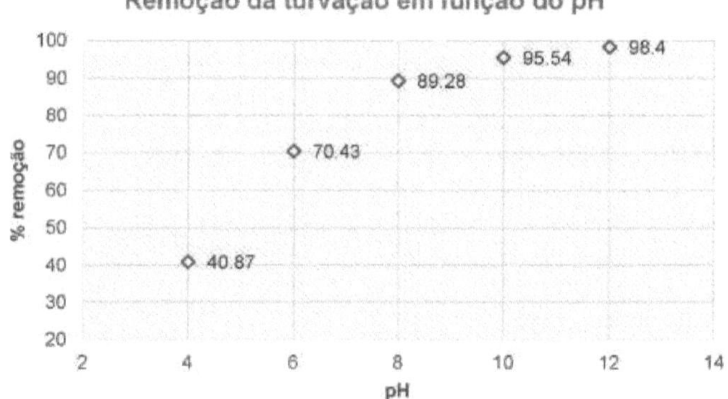

Figure 4.5: Effect of pH on the turbidity removal percentage

Table 4.7 and Figure 4.5 show that the use of the coagulant reduced turbidity from 345 NTU to 5.51 NTU, corresponding to 98.4% turbidity removal, with a dose of 240 mg/L and a pH of 12. At the end of stage 2, the final pH value was around the initial pH and there was a significant increase in the conductivity of the treated sample in all jugs as the pH increased. This variation in conductivity with increasing pH is due to the reagent used to adjust the initial pH of the sample.

Figure 4.5 shows that the percentage of turbidity removal increases progressively as the pH of the sample rises from 4 to 12.

Third stage of the jar test: Confirmation of the optimum dose of coagulant.

This stage served to confirm the optimum dose of coagulant in tests carried out at the optimum pH (12) found in stage 2. The sample submitted for treatment had an initial turbidity of 345 NTU, initial conductivity of 617 pS/cm and an initial (natural) pH of 5.58.

At the end of this stage, parameters were measured in the supernatant of the treated sample, and the values obtained are shown in Table 4.8.

Table 4.8: Effect of coagulant dose on turbidity removal at initial pH 12

Dose (mg)	Final pH	EC (pS/cm)	Turbidity (NTU)	% redemption
200	12,17	4270	12,4	96
220	12,27	4880	10,1	97
240	12,29	5000	6,7	98
260	12,27	5260	5,7	98
280	12,34	5720	9,3	97

Figure 4.6 shows the turbidity removal efficiency as the coagulant dose varies.

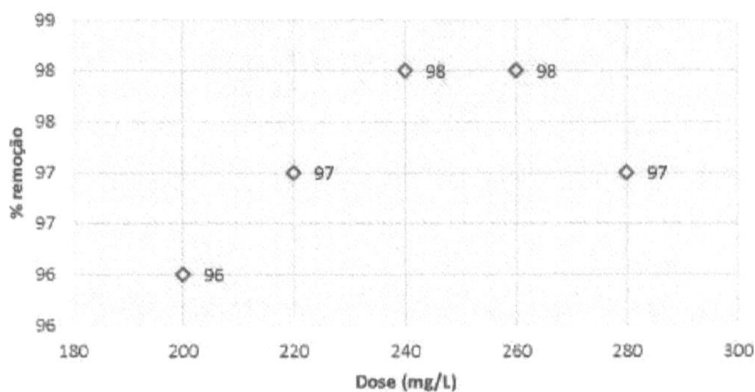

Figure 4.6: Effect of the coagulant dose on the turbidity removal percentage

The results in Table 4.8 and Figure 4.6 show that the coagulant reduced turbidity from 345 NTU to 5.7 NTU, corresponding to 98% turbidity removal, with a dose of 260 mg/L at an initial pH of 12. The final pH value was around the initial pH value (12) and the conductivity value increased significantly from 617 pS/cm to 5260 pS/cm. The possible reason for the significant increase in conductivity was mentioned earlier and is attributed to the saline extract of the coagulant and the reagent used (NaOH) to adjust the pH to 12.

It should be noted that the optimum dose of coagulant found in this last stage (260 mg/L) led to a final turbidity (5.7 NTU) slightly higher than that found in the previous stage (5.5 NTU) with the 240 mg/L dose.

4.2.2.2 Characterisation of the wastewater after the coagulation-flocculation test under optimum conditions

The raw wastewater was subjected to a coagulation-flocculation test using the optimum pH values (pH=12) and coagulant dose (260 mg/L) found in section 4.2.2.1. Table 4.9 shows the values of the parameters measured in the untreated wastewater and in the wastewater treated by the coagulation-flocculation process carried out under the optimum conditions.

Table 4.9: Characterisation of raw wastewater and wastewater treated with saline extract of *Moringa oleifera* Lam under optimum conditions.

Parameter	Waste water	Treated wastewater
pH	5,58	12,27
Conductivity (pS/cm)	617	5260
Turbidity (NTU)	345	5,7
COD (mgO2/L)	1296	1166
TSS (mg/L)	314	-

Pb (mg/L)	< LD	< LD	
Zn (mg/L)	0,0286		0,0276
Cu (mg/L)	< LD	< LD	
Cr (mg/L)	0,0207	< LD	
Fe (mg/L)	< LD	< LD	
Ni (mg/L)	< LD	< LD	

Table 4.9 shows that the coagulation-flocculation treatment helped to reduce turbidity by 98.4% and COD by 10%. There was a notable increase in conductivity, which can be attributed to the nature of the coagulant, which is a saline extract of a cationic protein, and to the reagent used to adjust the pH to 12. The treatment had no significant effect on the levels of heavy metals measured. These levels, which are below the emission limit values, were measured by atomic absorption spectrophotometry on acidified and digested samples as described in Chapter 3. The reduction in COD with treatment can be attributed to the reduction of suspended solids containing organic matter.

4.2.3 Coagulation-flocculation test with ground moringa seed without hulls

The powder from the shelled ground seed was obtained following the procedure reported by Price (2000) as described in Chapter 3.

4.2.3.1 Determining the optimum test conditions

The optimum conditions were determined in three stages, as mentioned above. Each stage consisted of coagulation, flocculation and sedimentation phases. At each stage, the pH, conductivity and turbidity values of the sample without coagulant were measured. At the end of each stage, after the sedimentation phase, these parameters were measured in the supernatant of each jug.

First stage of the jar test: Study of the effect of varying the coagulant dose. This stage was used to determine the optimum dose of coagulant for which the greatest turbidity removal occurred.

Before the jug test, some parameters were measured for this sample and the following results were obtained: natural pH 6.58, conductivity 1918 pS/cm and turbidity 378 NTU. The final turbidity value obtained and the turbidity removal percentage as a function of the coagulant dose are shown in Table 4.10 and Figure 4.7.

Table 4.10: Effect of coagulant dose on turbidity removal at the sample's natural pH (step 1)

Dose (mg/L)	pH	EC (pS/cm)	Turbidity (NTU)	% redemption
40	6,51	1907	317	16,1
60	6,52	1916	273	27,7
80	6,67	1917	189	50,0
100	6,62	1917	166	56,0
120	6,57	1917	118	68,7
140	6,55	1919	105	72,2
160	6,78	1919	332	12,1
180	6,55	1916	341	9,7
200	6,57	1917	158	58,2

Figure 4.7 shows the turbidity removal efficiency for different doses of coagulant.

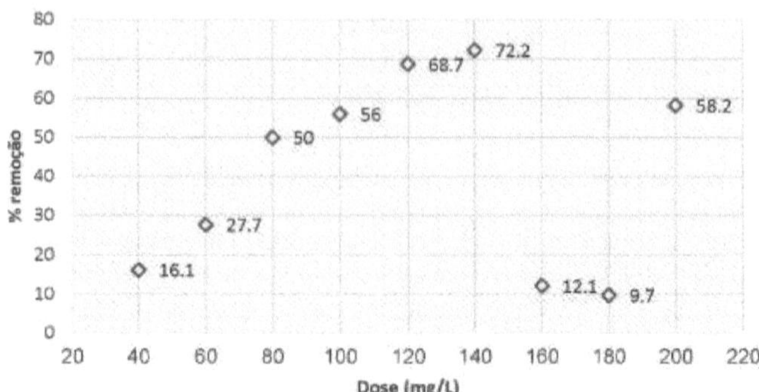

Figure 4.7: Effect of the coagulant dose on the turbidity removal percentage

From these results, it can be seen that there was a reduction in turbidity from 378 NTU to 105 NTU, corresponding to 72.2% turbidity removal, with the 140 mg/L dose. In each jug at the end of this stage, the final pH value of the sample was around the initial pH (6.5) and the final conductivity value was around 1916 pS/cm. The optimum dose found in stage 1 was therefore 140 mg/L, the dose at which the greatest turbidity removal occurred.

Second stage of the jug test: Study of the effect of varying the pH. The aim of this second stage was to find the pH value at which coagulation should be carried out for the greatest removal of turbidity. This pH was determined using the optimum dose of coagulant found in stage 1 (140 mg/L). In each jug, the initial pH value was varied between 4 and 12 in order to cover the alkaline and acidic pH ranges. The sample submitted for treatment had a natural pH of 6.58, conductivity of 1918 pS/cm and turbidity of 378 NTU.

The final turbidity values and the turbidity removal percentage as a function of the different initial pH values are shown in Table 4.11 and Figure 4.8.

Table 4.11: Effect of initial pH on turbidity removal with a coagulant dose of 140 mg/L (step 2)

KHB	Final pH	EC (pS/cm)	Turbidity (NTU)	% redemption
4	4,18	2500	271	28,3
6	6,03	1972	215	43,1
8	8,94	2080	101	73,2
10	10,14	2100	48,3	87,2
12	12,05	3350	13,2	96,5

Figure 4.8 shows the turbidity removal efficiency for different initial pH values of the sample.

Effectiveness of pH in removing turbidity

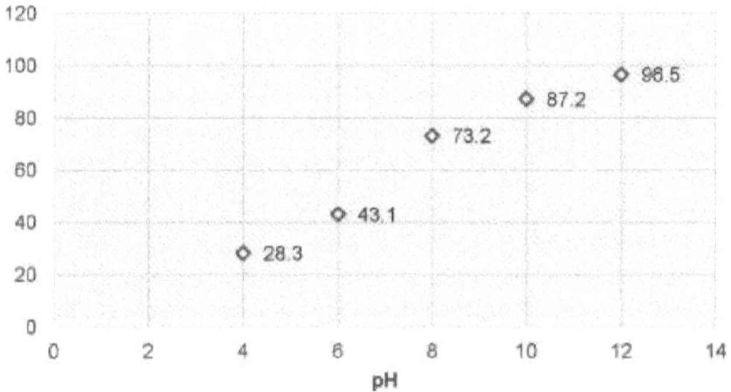

Figure 4.8: Effect of pH on the turbidity removal percentage

These results show that turbidity decreases with increasing pH, reaching a maximum value of 13.2 NTU at pH 12 (optimum pH), corresponding to 96.5% turbidity removal at a dose of 140 mg/L of coagulant. At the end of stage 2, the measured pH value was close to the initial pH. In the alkaline range, conductivity increased significantly with increasing pH.

Third stage of jar testing: Confirmation of the optimum dose of coagulant.

This stage served to confirm the optimum dose of coagulant in tests carried out at the optimum pH (12) found in stage 2. The sample submitted for treatment had an initial turbidity of 378 NTU, an initial conductivity of 1918 pS/cm and an initial (natural) pH of 6.58.

At the end of this stage, parameters were measured in the supernatant of the treated sample, and the values obtained are shown in Table 4.12.

Table 4.12: Effect of coagulant dose on turbidity removal at an initial pH of 12.

Dose (mg)	Final pH	EC (pS/cm)	Turbidity (NTU)	% redemption
120	11,37	3950	10,7	97
140	11,57	3930	12,5	97
160	11,66	3960	24,0	94

The turbidity removal efficiency with the coagulant dose is shown in Figure 4.9.

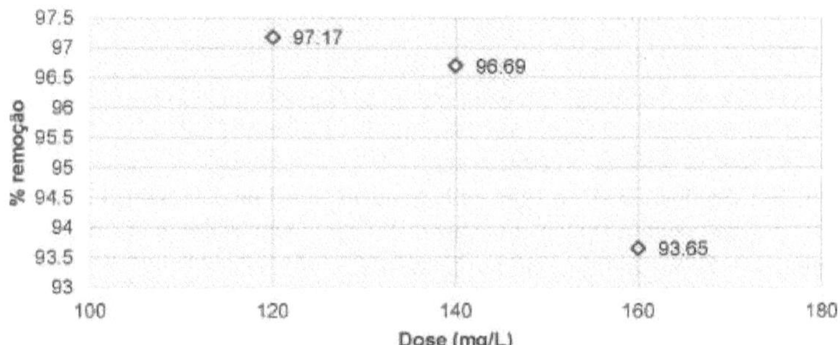

Figure 4.9: Effect of coagulant dose on turbidity removal percentage

The results obtained in this stage show that at a dose of 120 mg/L of coagulant, turbidity fell from 378 NTU to 10.7 NTU, corresponding to 97% turbidity removal. Compared to stage 2, a lower final turbidity was obtained with a lower dose of coagulant. The final pH value remained constant in all jugs at around the initial pH 12 and the final conductivity value varied greatly in relation to the initial conductivity value from 1918 pS/cm to 3950 pS/cm. This increase in conductivity can be attributed to the reagent added to the sample to adjust its pH to 12.

With this coagulant, as with the other coagulants tested, the results show once again that pH is a crucial parameter for achieving greater turbidity removal.

4.2.3.2 Characterisation of wastewater after coagulation test under optimum conditions

The raw wastewater was subjected to a coagulation-flocculation test using the optimum pH values (pH=12) and coagulant dose (120 mg/L) found in section 4.2.3.1. Table 4.13 shows the values of the parameters measured in the untreated wastewater and in the wastewater treated by the coagulation-flocculation process carried out under the optimum conditions.

Table 4.13: Characterisation of raw wastewater and wastewater treated with ground, shelled *Moringa oleifera* Lam seed under optimum conditions.

Parameter	Waste water	Treated wastewater
pH	6,58	11,66
Conductivity (pS/cm)	1918	3950
Turbidity (NTU)	378	10,7
COD (mgO2/L)	1296	497
TSS (mg/L)	328	-
Pb (mg/L)	< LD	< LD
Zn (mg/L)	0,0286	0,0288
Cu (mg/L)	< LD	< LD
Cr (mg/L)	0,0207	0,0305
Fe (mg/L)	< LD	< LD
Ni (mg/L)	< LD	< LD

The analysis in Table 4.13 shows that the coagulation-flocculation treatment helped to reduce turbidity by 97% and COD by 62%. However, there was a considerable increase in conductivity which can be attributed to the nature of the coagulant, which is a cationic protein, and to the reagent added to the sample to adjust its pH. The treatment had no significant effect on the levels of heavy metals measured. These levels, which are below the emission limit values, were measured by atomic absorption spectrophotometry on acidified and digested samples as described in Chapter 3. The reduction in COD with treatment can be attributed to the reduction of suspended solids containing organic matter.

4.2.4 Coagulation-flocculation test with ground seed and moringa hulls

The procedure for preparing ground seed powder with hulls is similar to that for ground seed without hulls mentioned in point 4.2.3.

4.2.4.1 Determining the optimum test conditions

The optimum conditions were determined in three stages, as mentioned above. Each stage consisted of coagulation, flocculation and sedimentation phases. At each stage, the pH, conductivity and turbidity values of the sample without coagulant were measured. At the end of each stage, after the sedimentation phase, these parameters were measured in the supernatant of each jug.

First stage of the jar test: Study of the effect of varying the coagulant dose. This stage was used to determine the optimum dose of coagulant for which the greatest turbidity removal occurred.

Before the jug test, some parameters were measured for this sample and the following results were obtained: natural pH 6.44, conductivity 1945 pS/cm and turbidity 355 NTU. The final turbidity value obtained and the turbidity removal percentage as a function of the coagulant dose are shown in Table 4.14 and Figure 4.10.

Table 4.14: Effect of coagulant dose on turbidity removal at the sample's natural pH (step 1)

Dose (mg/L)	pH	EC (pS/cm)	Turbidity (NTU)	% redemption
40	6,96	1945	269	24,2
60	6,97	1945	258	27,3
80	6,84	1946	274	22,8
100	6,85	1944	273	23,1
120	6,91	1936	239	32,6
140	6,90	1945	217	38,8
160	6,83	1941	193	45,6
180	6,69	1942	220	38,0
200	6,70	1952	185	47,9
220	6,76	1955	168	52,7
240	6,81	1947	229	35,5
260	6,92	1939	289	18,6
280	6,93	1949	236	33,5
300	6,85	1950	223	37,2
320	6,81	1944	208	41,4

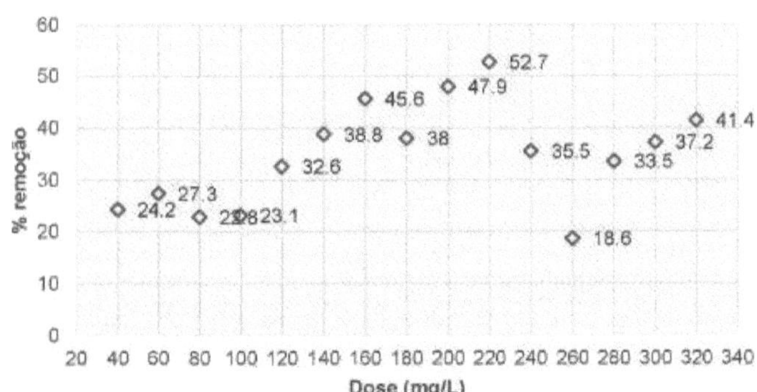

Figure 4.10: Effect of the coagulant dose on the turbidity removal percentage

These results show that the use of the coagulant reduced turbidity from 355 NTU to 168 NTU, corresponding to 52.7% turbidity removal, at a dose of 220 mg/L. For doses of coagulant lower or higher than 220 mg/L, the final turbidity obtained was higher than 168 NTU. The pH and conductivity values at the end of this stage were around the initial pH (6.4) and initial conductivity (1945 pS/cm) of the sample. In this case, the addition of the coagulant to the sample had no effect on the change in conductivity.

The optimum dose found in stage 1 was 220 mg/L, where the greatest turbidity removal occurred.

- Second stage of the jug test: Study of the effect of varying the pH. As already mentioned, the aim of this second stage was to find the pH value at which coagulation should be carried out for the greatest removal of turbidity. This pH was determined using the optimum dose of coagulant found in stage 1 (220 mg/L). The initial pH value of each jug was varied between 4 and 12 to cover the alkaline and acidic pH ranges. The sample submitted for treatment had a natural pH of 6.44, conductivity of 1945 pS/cm and turbidity of 355 NTU.

The final turbidity values and the turbidity removal percentage as a function of the different initial pH values are shown in Table 4.15 and Figure 4.11.

Table 4.15: Effect of initial pH on turbidity removal with a coagulant dose of 220 mg/L (step 2)

pH	Final pH	EC (pS/cm)	Turbidity (NTU)	% redemption
4	3,95	2500	347	2
6	6,07	2030	303	15
8	8,08	2190	128	64
10	9,77	2190	44,2	88
12	11,73	3710	30	92

The percentages of turbidity removal for each adjusted initial pH are shown in Figure 4.11.

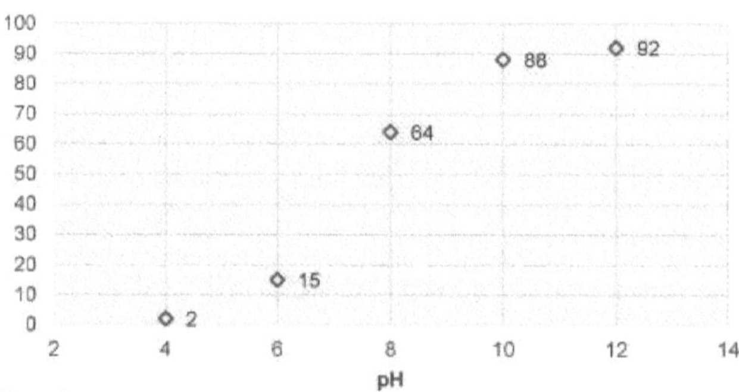

Figure 4.11: Effect of pH on the turbidity removal percentage

These results show that turbidity decreased progressively from 355 NTU to 30.0 NTU as the pH increased to a value of 12. At this optimum pH value, 92.0% turbidity removal was achieved with a coagulant dose of 220 mg/L. At the end of this stage, the measured pH value was always around the initial pH value and the conductivity value rose when compared to the initial value, with the greatest rise at pH 12. This rise can be attributed to the reagent added to adjust the pH.

Third stage of jar testing: Confirmation of the optimum dose of coagulant.

This stage served to confirm the optimum dose of coagulant in tests carried out at the optimum pH (12) found in stage 2. The sample submitted for treatment had an initial turbidity of 355 NTU, initial conductivity of 1945 pS/cm and an initial (natural) pH of 6.44.

At the end of this stage, parameters were measured in the supernatant of the treated sample, and the values obtained are shown in Table 4.16 and Figure 4.12.

Table 4.16: Effect of coagulant dose on turbidity removal at an initial pH of 12.

Dose (mg)	Final pH	EC (pS/cm)	Turbidity (NTU)	% redemption
180	11,82	3950	33,16	91
200	11,98	3930	32,30	91
220	11,90	3960	33,80	90
240	11,88	3950	33,60	91
260	11,87	3940	35,80	90

Figure 4.12 illustrates the turbidity removal efficiency as the coagulant dose varies.

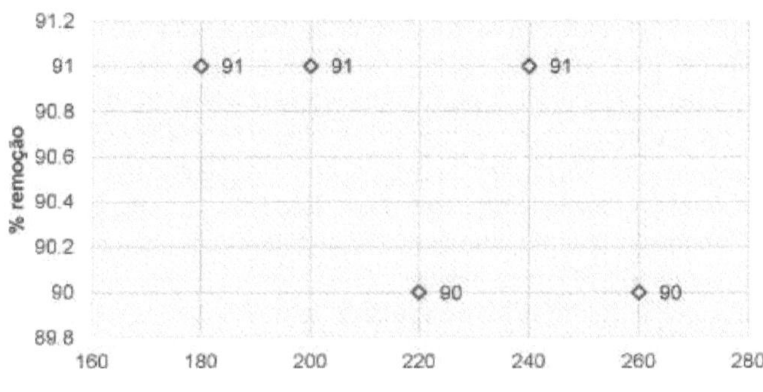

Figure 4.12: Effect of the coagulant dose on the turbidity removal percentage

The results obtained in this stage show that the use of the coagulant reduced turbidity from 355 NTU to 32.3 NTU, corresponding to 91% turbidity removal with the 200 mg/L dose. This turbidity removal is slightly lower than that obtained at the end of stage 2 with the 220 mg/L dose. The final pH value was around the initial pH 12 and the final conductivity value (3930 pS/cm) was much higher than the initial conductivity value (1945 pS/cm). This rise can be attributed to the reagent used to adjust the initial pH.

In stage 3, the optimum dose was found to be 200 mg/L because there was greater turbidity removal.

With this coagulant, as with the other coagulants tested, the results show once again that pH is a crucial parameter for achieving greater turbidity removal.

4.2.4.2 Wastewater characterisation after coagulation-flocculation test under optimum conditions

The raw wastewater was subjected to a coagulation-flocculation test using the optimum pH values (pH=12) and coagulant dose (200 mg/L) found in section 4.2.4.1. Table 4.17 shows the values of the parameters measured in the untreated wastewater and in the wastewater treated by the coagulation-flocculation process carried out under the optimum conditions.

Table 4.17: Characterisation of raw wastewater and wastewater treated with ground *Moringa oleifera*

Lam seed and bark under optimum conditions.

Parameter	Waste water	Treated wastewater
pH	6,44	11,82
Conductivity (pS/cm)	1945	3930
Turbidity (NTU)	355	32,30
COD (mgO2/L)	1296	812
TSS (mg/L)	328	-
Pb (mg/L)	< LD	< LD
Zn (mg/L)	0,0286	0,0475
Cu (mg/L)	< LD	0,0758
Cr (mg/L)	0,0207	< LD
Fe (mg/L)	< LD	< LD
Ni (mg/L)	< LD	< LD

Analysing Table 4.17 shows that treatment by coagulation-flocculation helps to reduce turbidity (90.9%) and COD (37%). However, there is a considerable increase in conductivity which can be attributed to the reagent added to the sample to adjust its initial pH. The treatment had no significant effect on the levels of heavy metals measured. These levels, which are below the emission limit values, were measured by atomic absorption spectrophotometry on acidified and digested samples as described in Chapter 3. The reduction in COD with treatment can be attributed to the reduction of suspended solids containing organic matter.

4.2.5 Comparison of the different coagulants tested

After the different tests with the natural coagulants (aqueous extract, saline extract, ground seeds without hulls and ground seeds with hulls), the turbidity removal efficiency was compared to determine the best coagulant.

Table 4.18 shows the optimum pH values and coagulant dose, and the different turbidity and organic matter removal percentages obtained with the four coagulants used in wastewater treatment.

Table 4.18: Optimum pH values, optimum coagulant dose and percentage removal of turbidity and organic matter with the coagulants tested

Coagulant	Optimum pH	Optimum dose - (mg/L)	(%) redemption	
			COD	Turvagao
Aqueous extract	12	260	29	95
Salt extract	12	260	10	98
Ground shelled seeds	12	120	62	97
Ground seed with shell	12	200	37	91

It can be seen that the salt extract coagulant had a higher turbidity removal capacity of 98% at a high dose of 260 mg/L. In comparison, the hulled ground seed coagulant was able to remove slightly less turbidity (97%) at a lower dose of 120 mg/L. In fact, it can be concluded that the best coagulant tested was shelled ground seed, because of its turbidity (97%) and COD (62%) removal power at a lower dose than the other coagulants. It should be noted that all the coagulants increased the conductivity of the treated wastewater.

According to Madrona (2010), in his study on the treatment of surface water with coagulants (aqueous extract and saline extracts (0.01, 0.1 and 1 M NaCl)), found that waters with a turbidity of 550 NTU achieve turbidity removal percentages of over 90% when an optimum dose of 50 mg/L is used and that for waters with a turbidity of 150 NTU, the removal percentages are in the 20% range for an optimum coagulant dose of 25 mg/L.

The work carried out by Vieira *et al.* (2010) cited by Silva *et al.* (2018) on the use of Moringa seed powder as a biosorbent in the removal of organic compounds from an effluent from the dairy industry led to good results. The treatment of an effluent with an organic matter load of up to 1.0 g/L was carried out using 0.2 g of Moringa seed and this mass removed approximately 98% of the colour and turbidity present in the sample.

The team of Pereira *et al.* (2011) carried out coagulation-flocculation tests to treat a wastewater containing oils and fats with saline and aqueous extracts of *Moringa oleifera* Lam and reported that the saline extract removed 96% of the oil from the water with only 5.0 mg/L.

Schmitt (2011) achieved a similar result, obtaining a 93 per cent removal rate with the use of saline extract from *Moringa oleifera* seeds in the treatment of wastewater from the dairy industry.

The percentage of turbidity removal found in this work with the coagulants tested is similar to that reported in the literature.

4.3 Organic matter adsorption tests

The two adsorbents, *Moringa oleifera* Lam seed husk and pod, were crushed and subjected to the methodology described in section 3.5 of Chapter 3 for the adsorption of organic matter and heavy metals. Organic matter in soluble form can be removed by adsorption.

The efficiency of the adsorbents for the adsorption of organic matter and metals was evaluated by varying the pH, contact time and mass of adsorbent in experiments carried out in decontamination mode.

4.3.1 Adsorption test with seed husk as adsorbent

4.3.1.1 Determination of PCZ and effect of pH on adsorption

Evaluating the pH or point of zero charge (PCZ) aims to gain a better understanding of the surface electrical behaviour of solids used as adsorbents. Experiments were carried out to determine the pH at which the surface charges cancel each other out. As the charge of the organic material is not known, the best pH for adsorption cannot be predicted from the value found for the PCZ. The results obtained are shown in Table 4.19 and Figure 4.13.

Table 4.19 reports the different pH values measured after contacting the seed husk with 0.5 and 0.05 mol/L KCl solutions initially adjusted to pH values from 2 to 10.

Table 4.19: Determination of the Zero Load Point of the Moringa oleifera Lam seed husk

	Final pH		ApH = final pH - initial pH	
Initial pH	0.5 mol/ L0	.05 mol/L	0.5 mol/L	0.05 mol/L
2	2,49	2,34	0,49	0,34
4	4,67	4,45	0,67	0,45
6	7,12	7,30	1,12	1,30
8	7,43	7,30	-0,57	-0,70
10	7,71	7,44	-2,29	-2,56

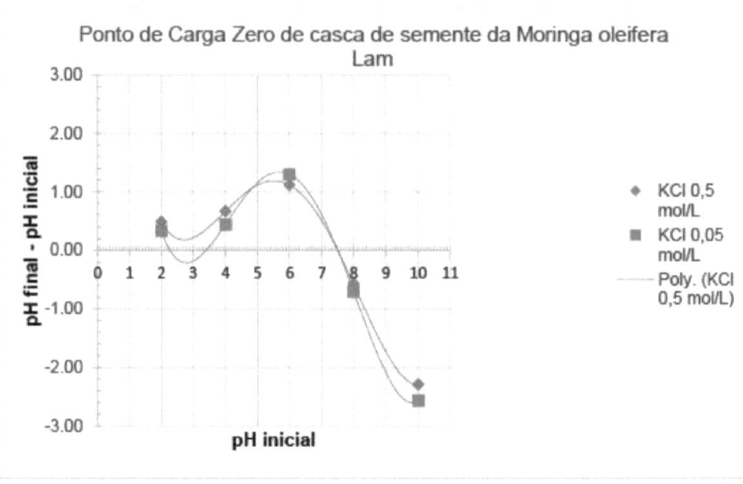

Figure 4.13: Determining the Zero Load Point of the seed coat

It can be seen from Figure 4.13 that the PCZ found with both the 0.5 and 0.05 mol/L KCl solutions is 7.5. This means that the adsorbent has a negative surface charge for initial pH values greater than 7.5, indicating that the adsorption of cations is favoured for initial pH values greater than 7.5.

For initial pH values lower than 7.5, this adsorbent is positively charged and will adsorb more negative species.

An adsorption study was carried out by varying the initial pH by different values (2, 4, 6, 8 and 10) to determine the pH at which the adsorbent has the capacity to adsorb the greatest amount of organic matter. The sample studied had an initial COD value of 418 $_{mgO2/L}$.

Table 4.20 shows the effect of pH on the adsorption of organic matter and its removal percentage.

Table 4.20: Effect of pH on the removal of organic matter (COD)

pH	COD ($_{mgO2/L}$)	% redemption
2	185	56
4	200	52
6	252	40
7,5	365	13
8	231	45
10	241	43

Table 4.20 shows that the greatest removal of organic matter (56%) was obtained at a pH of 2. This was the pH chosen for the kinetic and equilibrium studies of organic matter adsorption.

Figure 4.14 illustrates the percentages of organic matter removal at each initial pH value.

Effect of pH on the adsorption of organic matter

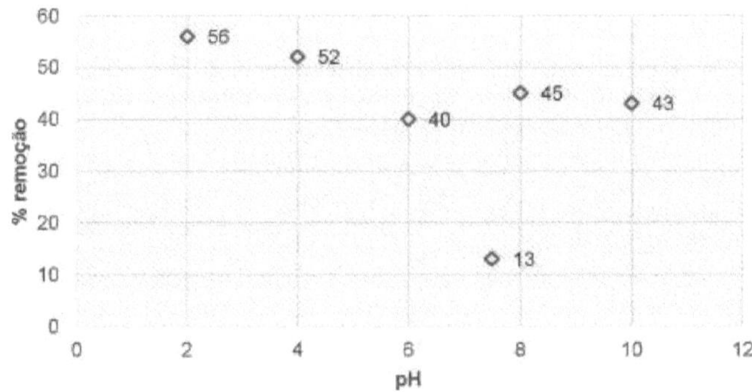

Figure 4.14: Effect of pH on the adsorption of organic matter with seed husks

Figure 4.14 shows that the highest removal of organic matter was 56% at pH 2. It can be seen that at the pH corresponding to the zero charge point of this adsorbent (7.5), the removal of organic matter was the lowest (13%).

4.3.1.2 Determination of adsorption kinetics

Based on the tests carried out as described in chapter 3, the concentration of organic matter in the solution was measured for different contact times (c_t) and from this, using equation (2.11), the concentration of organic matter adsorbed by the solid was calculated for the different contact times (q_t).

Tables 4.21 and 4.22 show the concentration of organic matter adsorbed as a function of contact time with the seed husk adsorbent at temperatures of 20°C and 40°C.

Table 4.21: Organic matter adsorbed by seed husk at 20 °C

Mass 0.05 g; Volume 0.1 L; C_0 415 mg /L; T 20 °C			
Adsorbent Time (minutes)	Ct (mg/L)	qt (mg/g)	% remogao
0	415	0	0,0
30	352	126	15,2
Casca60	322	186	22,4
120	322	186	22,4
180	324	182	21,9

Table 4.22: Organic matter adsorbed by seed husk at 40 °C

Mass 0.05 g; Volume 0.1 L; c_0 405 mg/L; T 40 °C			
Adsorbent Time (minutes)	Ct (mg/L)	qt (mg/g)	% redemption
0	405	0	0,0
30	343	124	15,3
Casca60	306	198	24,4
120	301	208	25,7
180	324	162	20,0

The results shown in Tables 4.21 and 4.22 indicate that the time taken to reach equilibrium is the same at both temperatures (60 min), with the amount of organic matter adsorbed being slightly higher at 40 °C.

The pseudo-first order and pseudo-second order kinetic models described in Chapter 2 by equations (2.5) to (2.10) were used to represent the amount of organic matter adsorbed by the solid for the different contact times.

The pseudo-first order and pseudo-second order models were fitted to the data at 20 and 40 °C using the Excel solver tool, minimising the sum of the squares of the deviations between the experimental and calculated qt values (SQD). The fitting results with the best parameter values found for each model are illustrated in Figures 4.15 and 4.16.

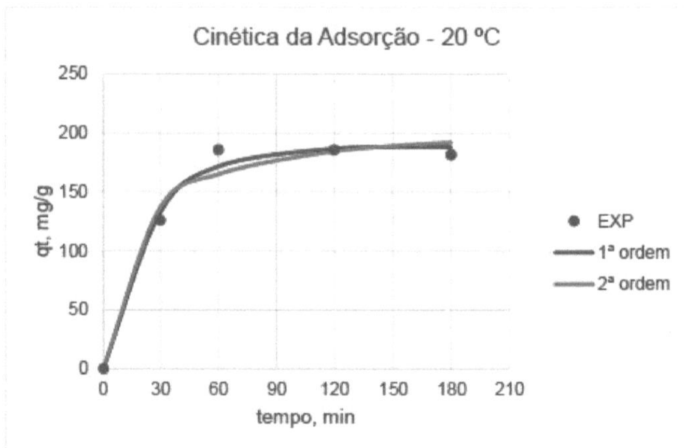

Figure 4.15: Fit of the pseudo-first and pseudo-second order models to the kinetic data obtained with the shell at 20 °C

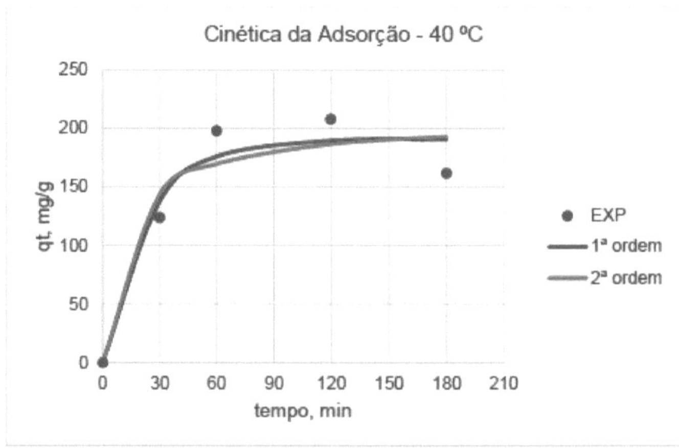

Figure 4.i6: Fit of the pseudo-first and pseudo-second order models to the kinetic data obtained with bark at 40 °C

Figures 4.i5 and 4.i6 show that the experimental value of the concentration of the adsorbed solute increases with contact time and from 60 minutes onwards a plateau is reached, showing that adsorption equilibrium has been reached.

Table 4.23 shows the parameter values of the two kinetic models that resulted in the best fit of these models to the experimental values of the concentration adsorbed by the solid. This table also shows the correlation coefficient (R) between the experimental values and those calculated by the model.

Table 4.23: Best parameter values of the kinetic models fitted to the adsorption of organic matter by moringa seed husk

Kinetic model						
Adsorbent	Pseudo-first order				Pseudo-second order	
T °C	q_e K_f SQD				q_e KsSQD	
	(mg/g) (min^{-1})R				(mg/g) (g/mg.min) R	
20 Bark	188, 40, 0404313		,7		209, 40,0003693 ,4	
	(0,994)				(0,987)	
40	190, 80,0431831 ,6				207, 40, 00042607,3	
	(0,967)				(0,953)	

(): value of R

Table 4.23 and Figures 4.15 and 4.16 show that the pseudo-first order model is the one that best fits the experimental data obtained at the two temperatures of 20 and 40 °C (lower SQD values and higher R values). It can also be seen that the calculated values of q_e for the pseudo-first order model are closer to the experimental values obtained at the plateau and that temperature has no significant effect on adsorption kinetics.

Souza (2016) tested three adsorbents (Moringa oleifera Lam seed, bark and pod) in the biosorption process for the removal of Diuron from contaminated water, proving that the pseudo-second order model fitted the experimental data best. The correlation coefficient of the pseudo-second order model was higher than that of the pseudo-first order model for the three adsorbents studied and the estimated value of q_e for this model was also closer to the experimental value.

Febrianto *et al.* (2009) compared the pseudo-first order kinetic model with the pseudo-second order kinetic model to represent kinetic data in biosorption systems and concluded that the pseudo-second order model is considered more suitable for representing this type of data.

4.3.1.3 Determining the adsorption equilibrium

Based on the tests carried out as described in chapter 3, the concentration of organic matter in the solution at the end of the equilibrium time was measured for different adsorbent masses (c_e) and from this, using equation (2.11), the concentration of organic matter adsorbed by the solid at equilibrium was calculated for the different adsorbent masses (q_e).

Tables 4.24 and 4.25 show the c_e and q_e values for temperatures of 20°C and 40°C, respectively.

Table 4.24: Concentration values of organic matter at equilibrium at 20 °C

C_0 378 mg/L; Volume 0,'		L; T 20 °C; 150 rpm; time 60 min		
AdsorbentMass (g)C_e		(mg/L)	q_e (mg/g)	% Removal
0,03		341	123	10
0,05		288	180	24

0,1	288	90	24
0,15	278	67	26
Bark0 ,2	142	118	62
0,3	37	114	90
0,4	100	70	74
0,5	58	64	85
0,6	58	53	85

Table 4.25: Concentration values of organic matter at equilibrium at 40 °C

C0 378 mg/L; Volume 0.1 L; T 40 °C; 150 rpm; time 60 min			
AdsorbentMass (g)	C_e (mg/L)	q_e (mg/g)	% Removal
0,02	307	355	19
0,03	292	287	23
0,04	277	253	27
0,05	224	308	41
Bark0 ,08	181	246	52
0,1	116	262	69
0,15	86	195	77
0,2	60	159	84
0,3	45	111	88
0,4	45	83	88
0,5	15	73	96

Tables 4.24 and 4.25 show that at 20 °C, with a mass of 0.3 g, there was the greatest removal of organic matter (90 per cent), while at 40 °C, the highest percentage of removal (96 per cent) was obtained with a mass of 0.5 g.

The Langmuir and Freundlich equiKbrium models described in Chapter 2 by equations (2.12) to (2.18) were used to represent the experimental equiKbrium data.

Figures 4.17 and 4.18 show the best fit of the Langmuir and Freundlich models to the experimental data. The parameter values of these two equilibrium models were obtained by fitting the respective non-linear equations to the experimental equilibrium data, minimising the sum of the squares of the deviations (SQD) between the experimental and calculated values of q_e, using the excel solver.

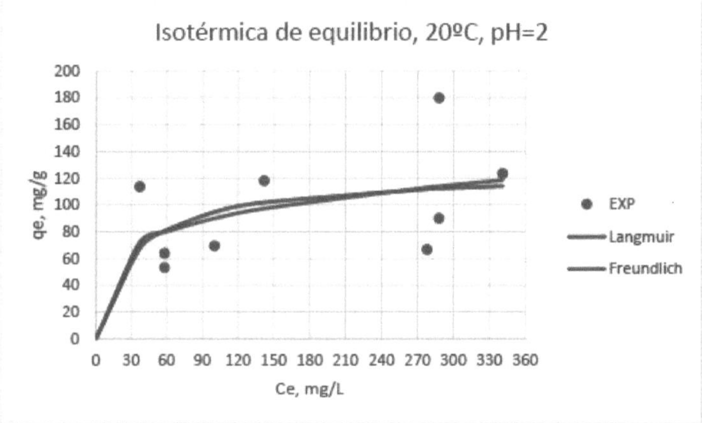

Figure 4.17: Fit of the Langmuir and Freundlich models to the experimental adsorption equilibrium data obtained at 20 °C with moringa bark

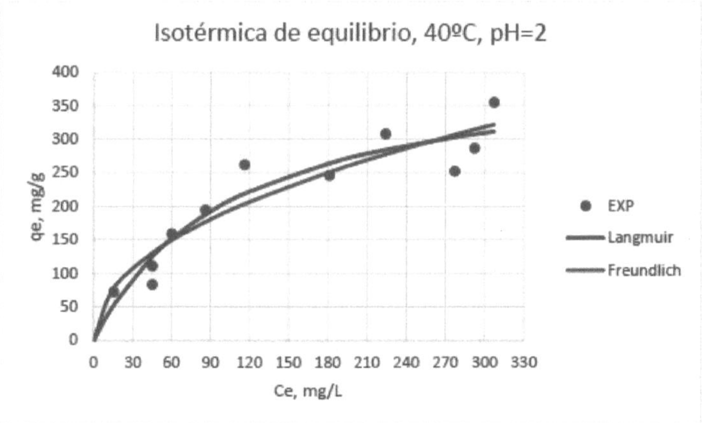

Figure 4.18: Fit of Langmuir and Freundlich models to experimental adsorption equilibrium data obtained at 40 °C with moringa bark

The parameter values obtained by the best fit of the theoretical Langmuir and Freundlich models to the experimental data are shown in Table 4.26. This table also shows the correlation coefficient (R) between the experimental values and those calculated by the model.

Table 4.26: Parameter values obtained by the best fit of the Langmuir and Freundlich equilibrium models to the experimental adsorption data using moringa bark

Equilibrium model						
AdsorbentLangmuirFreundlich						
		T q_{max}		KlSQD	Kfn SQD^{-1}	
°C						
(mg/g)	(L/mg) R	(mg/g)/(mg/L)^{A} (1/n)		R		
20124,	40	,03211288,	732,	10,	2210630,8	
(0,370)	(0,429)					
Casca 40418,	50	,009510071,	421,	70,	4712199,1	
(0,942)	(0,930)					

(): value of R

Table 4.26 and Figures 4.17 and 4.18 show that the Freundlich model is the one that best fits the experimental data at 20 °C.
best fits the experimental data at 20 °C, although the correlation between experimental and calculated values is moderate (0.40<R<0.69), while the
Langmuir model fits the data better at 40 °C. At this temperature, the correlations between experimental and calculated values for both models are very strong (R>0.90). In the Langmuir model, the maximum adsorption capacity increases with increasing temperature and the equilibrium constant decreases with increasing temperature, indicating that the process is exothermic. In the Freundlich model, the adsorption capacity (K_F) and the heterogeneity of the adsorbent surface (1/n) decrease with increasing temperature. The closer 1/n is to zero, the greater the heterogeneity of the adsorbent surface.
Souza (2016) used the seed, bark and pod of *Moringa oleifera* Lam in the biosorption process to remove Diuron from contaminated water and used the Langmuir and Freundlich equilibrium models. He concluded that the Freundlich model provided the best fit for the three adsorbents used, since these three adsorbents are very heterogeneous materials.

4.3.2 Pod adsorption test

4.3.2.1 Determination of PCZ and effect of pH on adsorption

The determination of the PCZ of the moringa pod is illustrated in Table 4.27 and Figure 4.19.

Table 4.27 shows the different pH values measured after contacting the *Moringa oleifera* Lam pod with 0.5 and 0.05 mol/L KCl solutions adjusted to initial pH values of 2 to 10.

Table 4.27: Determining the Zero Load Point of the Moringa oleifera Lam pod

Initial pH	Final pH		ApH = final pH - initial pH	
	0.5 mol/ L0	.05 mol/L	0.5 mol/L	0.05 mol/L
2	2,37	2,28	0,37	0,28
4	4,38	4,26	0,38	0,26
6	7,03	6,30	1,03	0,30
8	6,81	7,23	-1,19	-0,77
10	7,57	7,37	-2,43	-2,63

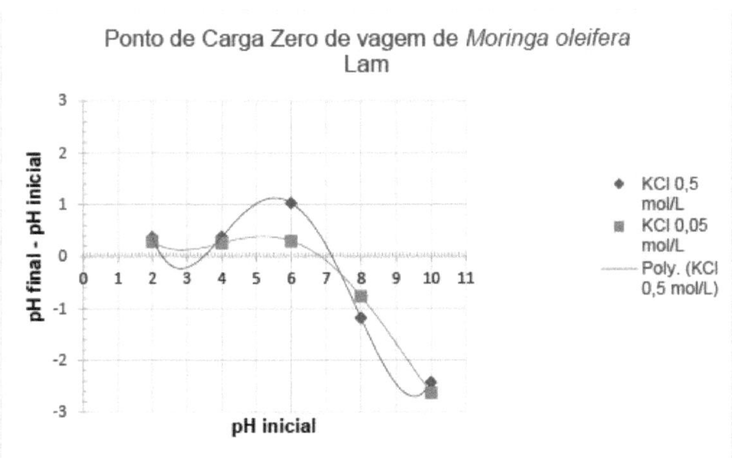

Figure 4.19: Determining the Zero Load Point of the Moringa oleifera Lam pod

Figure 4.19 shows that the PCZ found with the 0.5 mol/L KCl solution is 7.2, implying that the adsorbent has a negative surface charge for initial pH values higher than 7.2. This indicates that the adsorption of cations is favoured for initial pH values higher than 7.2. Conversely, for initial pH values below 7.2, the adsorbent becomes positively charged and will adsorb more negative species.

After determining the PCZ, an adsorption study of the organic material was carried out by varying the initial pH by different values (2, 4, 6, 7.2, 8 and 10) to determine the pH at which the adsorbent has the capacity to adsorb the greatest amount of organic material. The sample studied had an organic matter content translated into an initial COD value of 418 mgO2/L.

Table 4.28 shows the effect of pH on the adsorption of organic matter and its removal percentage.

Table 4.28: Effect of pH on the removal of organic matter (COD)

pH	COD (mgO2/L)	% redemption
2	230	45
4	270	35
6	270	35
7,2	306	27
8	332	21
10	337	19

Table 4.28 shows that there was greater removal of organic matter (45%) at pH 2. Removal decreases as the pH increases.

For the kinetic and equilibrium study of the adsorption of organic matter with this adsorbent, it was decided to use a pH of 6 because the removal of organic matter (35%) was closer to that obtained with a pH of 2 (45%).

Figure 4.20 shows the percentages of organic matter removal at each initial pH value.

Figure 4.20: Influence of pH on the removal of organic matter with pods

Figure 4.20 shows a tendency for the removal of organic matter to decrease as the pH increases.

4.3.2.2 Determination of adsorption kinetics

The adsorption kinetics were determined using the tests carried out as described in Chapter 3, measuring the concentration of organic material in the solution after different contact times (c_t) and using equation (2.11) to calculate the concentration of organic material adsorbed by the solid for the different contact times (q_t).

Tables 4.29 and 4.30 show the concentration of organic matter adsorbed as a function of contact time with the *Moringa oleifera* Lam pod adsorbent at temperatures of 20°C and 40°C.

Table 4.29: Organic matter adsorbed by the pod at 20 °C

Mass 0.05 g; Volume 0.1 L; C_o		445 mg/L; T 20 °C		
Adsorbent Time (minutes)	Ct (mg/L)	qt (mg/g)	%	remogao
0	445	0	0	0,0
30	363	164		18,4
Bean60	323	244		27,4
120	328	234		26,3
180	361	168		18,9

Table 4.30: Organic matter adsorbed by the pod at 40 °C

	Mass 0.05 g; Volume 0.1 L; C_o		363 mg/L; T 40 °C		
	Adsorbent Time (minutes)	Ct (mg/L)	qt (mg/g)	%	remogao
	0	363		0	0,0
	30	323		80	11,0
Pods	60	302		122	16,8
	120	292		142	19,6
	180	346		34	4,7

The results at 20 °C suggest that the adsorption equilibrium is reached at a contact time of 60 minutes. The results at 40 °C show a decrease in the amount of organic material adsorbed with increasing temperature and an unexpected variation with contact time, which could be due to the heterogeneity of the adsorbent.

Figures 4.21 and 4.22 show the fit of the pseudo-first order and pseudo-second order models to the experimental data at 20 and 40 °C.

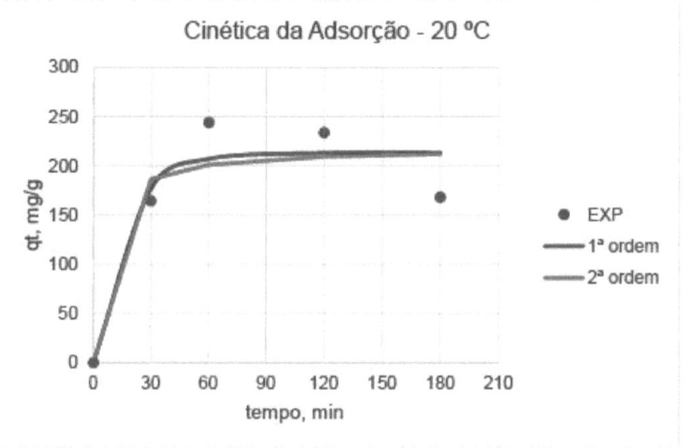

Figure 4.21: Fitting the pseudo-first order and pseudo-second order models to the kinetic data kinetic data obtained with pods at 20 °C.

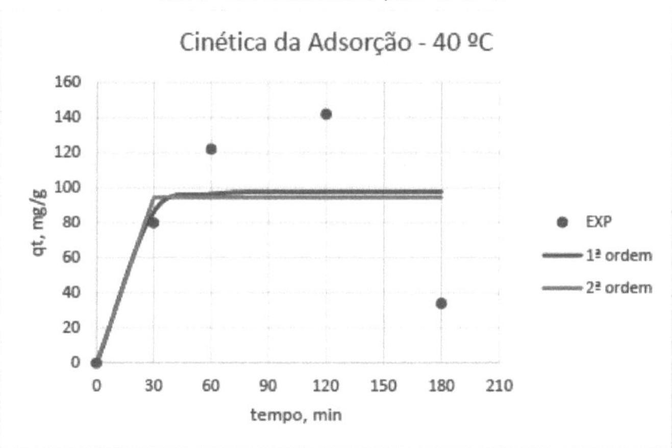

Figure 4.22: Fitting the pseudo-first order and pseudo-second order models to the kinetic data kinetic data obtained with the pod at 40 °C

Table 4.31 shows the parameter values of the two kinetic models that resulted in the best fit of these models to the experimental values of the concentration adsorbed by the solid. This table also shows the correlation coefficient (R) between the experimental values and those calculated by the model.

Table 4.31: Best parameter values of the kinetic models fitted to the adsorption of organic matter by moringa pods

Kinetic model		
Adsorbent	Pseudo-first order	Pseudo-second order

	T	q_e	K_f	SQD	q_e	K_s	SQD
	°C	(mg/g)	$(min)^{-1}$	R	(mg/g)	(g/mg.min)	R
Pods	20	213,5	0,059	4016,3	218,4	0,00091	4908,1
				(0,946)			(0,934)
	40	97,8	0,071	6718,3	94,5	0,464	6883,2
				(0,722)			(0,714)

(): value of R

Table 4.31 and Figures 4.21 and 4.22 show that based on the SQD and R values, the pseudo-first order model is the one that best fits the experimental data at both temperatures, although the difference between the models is not significant given the dispersion of the experimental points. At 20 °C the correlations between experimental values and those calculated by the two models are very strong (R>0.90). There is also an increase in the kinetic constants of the two models with increasing temperature.

4.3.2.3 Determining the adsorption equilibrium

The adsorption equilibrium was determined using the tests carried out as described in Chapter 3, measuring the concentration of organic matter in the solution at the end of the equilibrium time for different adsorbent masses (c_e) and using equation (2.11) to calculate the concentration of organic matter adsorbed by the solid at equilibrium for the different adsorbent masses (q_e).

Tables 4.32 and 4.33 show the C_e and q_e values for temperatures of 20°C and 40°C respectively.

Table 4.32: Concentration values of organic matter at equilibrium at 20 °C

C_0 428 mg/L; Volume (0.1 L); T 20 °C; 150 rpm; time 60 min			
Adsorbent Mass (g)	C_e (mg/L)	q_e (mg/g)	% redemption
0,05	337	182	21
0,1	307	121	28
0,15	287	94	33
Pods0 ,2	269	80	37
0,3	86	114	80
0,5	116	62	73
1	76	35	82

Table 4.33: Organic matter concentration values at equilibrium at 40 °C

	C_0 318m	g/L; Volume (0.1 L); T 40 °C		; 150 rpm; time 60 min	
Adsorbent		Mass (g)	C_e (mg/L)	q_e (mg/g)	% redemption
		0,02	203	575	36
		0,03	193	417	39
		0,04	193	313	39
		0,05	177	282	42
		0,06	172	243	46
		0,08	146	215	54
Pods		0,1	141	177	56
		0,15	130	125	59

0,2	130	94	59
0,3	36	94	89
0,4	31	72	90
0,5	57	52	82

Tables 4.32 and 4.33 show the decrease in the concentration of organic matter in the C_e solution (mg/L) and the consequent increase in its percentage of removal from the solution as the mass of the adsorbent increases up to a certain value. After a certain value of adsorbent mass, the adsorbent may lose its efficiency because its surface area is not fully available. The likelihood of this occurring is greater when the adsorbent is used in sachets, as is the case in this study.

Figures 4.23 and 4.24 show the best fit of the Langmuir and Freundlich models to the experimental data at 20 and 40 °C, respectively.

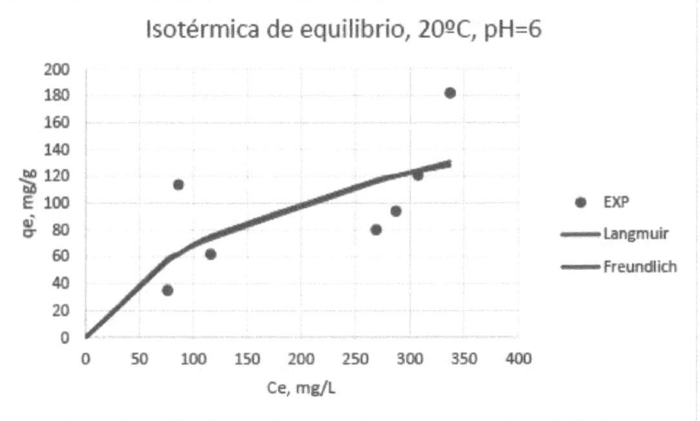

Figure 4.23: Fit of Langmuir and Freundlich models to experimental adsorption equilibrium data obtained at 20 °C with moringa pods

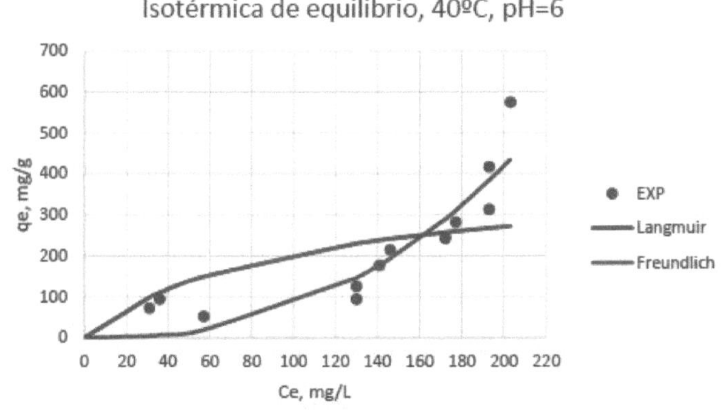

Figure 4.24: Fit of Langmuir and Freundlich models to experimental adsorption equilibrium data obtained at 40 °C with moringa pods

The parameter values obtained by the best fit of the theoretical models of

Langmuir and Freundlich to the experimental data are shown in Table 4.34.

Table 4.34: Parameter values obtained by the best fit of the
Langmuir and Freundlich equilibrium models to the experimental adsorption data using moringa pods

		Equilibrium model					
		Langmuir			Freundlich		
Adsorbent	T °C	q_{max} (mg/g)	K_L (L/mg)	SQD R	K_F (mg/g)/(mg/L)^A (1/n)	n^{-1}	SQD R
Pods	20	203,1	0,0051	8363,5 (0,617)	5,559	0,543	7920,0 (0,642)
	40	400	0,0104	161606,1 (0,704)	0,0009	2,448	45795,5 (0,918)

(): value of R

Table 4.34 and Figures 4.23 and 4.24 show that the Freundlich model is the one that best fits the experimental equilibrium data at temperatures of 20 and 40 °C, with the correlation between experimental and calculated values being very strong (R>0.90) only at 40 °C. In the Freundlich model, the value of 1/n that produces the best fit for the temperature of 40 °C is outside the range of values (between 0 and 1) reported for this parameter.

4.3.3 Comparison of the adsorbents tested

The parameters obtained by fitting the pseudo-first order and pseudo-second order kinetic models and the Langmuir and Freundlich isotherms to the experimental data obtained with the two adsorbents tested are compiled in Table 4.35 and 4.36, respectively.

Table 4.35: Parameters obtained by the best fit of the pseudo-first order
and pseudo-second order kinetic models to the experimental data

		Kinetic model					
Adsorbent		Pseudo-first order			Pseudo-second order		
	T °C	q_e (mg/g)	K_f (min^{-1})	SQD R	q_e (mg/g)	K_s (g/mg.min)	SQD R
Bark	20	188, (0,994)	40,0404313	,7	209, (0,987)	40,0003693	,4
	40	190, (0,967)	80,0431831	,6	207, (0,953)	40,00042607	,3
Pods	20	213, (0,946)	50,0594016	,3	218, (0,934)	40,00094908	,1
	40	97 (0,722)	80,0716718	,3	94, (0,714)	50,4646883	,2

(): value of R

Table 4.36: Parameters obtained by the best fit of the Langmuir and Freundlich equiKbrium models to
the experimental data

		Equilibrium model					
		Langmuir			Freundlich		
Adsorbent	T °C	q_{max} (mg/g)	K_L (L/mg)	SQD R	K_F (mg/g)/(mg/L)^A (1/n)	n^{-1}	SQD R
Bark	20	124,4	0,032	11288,7 (0,370)	32,1	0,223	10630,8 (0,429)
	40	418,5	0,0095	10071,4 (0,942)	21,7	0,471	12199,1 (0,930)
		203,1	0,0051	8363,5	5,559	0,543	7920,0

Pods	20			(0,617)			(0,642)
		400	0,0104	161606,1	0,0009	2,448	45795,5
	40			(0,704)			(0,918)

(): value of R

Analysing Tables 4.35 and 4.36 we can see that most of the results indicate that the adsorption kinetics are best represented by the pseudo-first order model and the adsorption equilibrium is best represented by the Freundlich model for both adsorbents. The results also indicate that the best adsorbent tested is *Moringa oleifera* Lam seed husk because it has a higher adsorption capacity and less heterogeneity.

4.4 Lead (Pb) adsorption tests

This section presents the results of the adsorption of lead (Pb) dissolved in wastewater using *Moringa oleifera* Lam seed husk as an adsorbent. The raw wastewater sample was brought to pH 12 to eliminate suspended solids. The supernatant of this treated water was adjusted to pH 2 and after adjusting the pH, lead was added to the sample. This is the pH value for the greatest removal of organic matter as shown in Table 4.20.

4.4.1 Determination of adsorption kinetics

From the tests carried out as described in chapter 3, the concentration of Pb in the solution was measured for different contact times (c_t) and from this, using equation (2.11), the concentration of Pb adsorbed by the solid was calculated for the different contact times (q_t).

Table 4.37 shows the concentration of Pb adsorbed as a function of contact time with the seed husk adsorbent at a temperature of 20°C. It should be noted that Pb was determined by EEA.

Table 4.37: Lead adsorbed by seed husk at 20 °C

C_0 11.5 mg/L; Volume 0.1 L; Mass 0.05 g; T 20 °C			
Adsorbent Time (minutes)	Ct (mg/L)	qt (mg/g)	% redemption
10	6,4	10,1	44,0
20	7,5	8,0	34,8
30	5,2	12,5	54,7
Bark 40	5,6	11,9	51,6
50	4,9	13,1	57,1
60	6,6	9,8	42,6

The results shown in Table 4.37 indicate that 60 minutes is a sufficient time for the adsorption equilibrium to be reached.

The pseudo-first order and pseudo-second order kinetic models described in Chapter 2 by equations (2.5) to (2.10) were used to represent the amount of Pb adsorbed by the solid for the different contact times.

The pseudo-first order and pseudo-second order models were fitted to the data at 20 °C using the Excel solver tool, minimising the sum of the squares of the deviations between the experimental and calculated qt values (SQD). The fitting results with the best parameter values found for each model are illustrated in Figure 4.25.

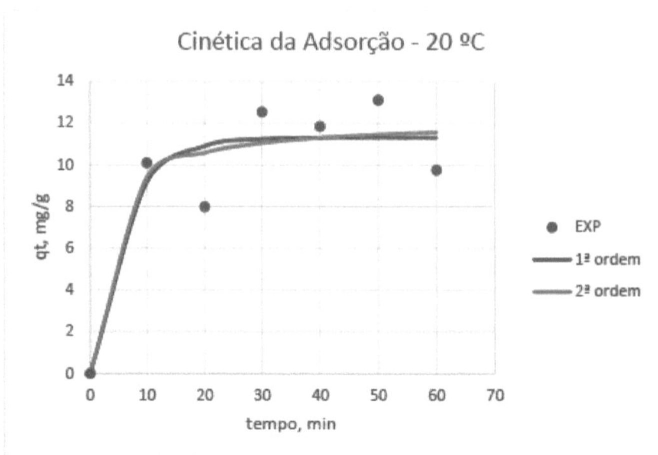

Figure 4.25: Fitting the pseudo-first-order and pseudo-second-order models to the kinetic data. kinetic Pb adsorption data obtained with bark at 20 °C

Figure 4.25 shows the formation of a plateau after 40 minutes, showing that the adsorption equilibrium has been reached.

Table 4.38 shows the best values found for the parameters of the two kinetic models and the corresponding SQD and R values.

Table 4.38: Best parameter values of the kinetic models fitted to Pb adsorption by moringa seed husk

Adsorbs nte		Kinetic model						
		Pseudo-first order			Pseudo-second order			
	T	q_e	K_f	SQD	q_e		K_S	SQD
	°C	(mg/g)	(min)$^{-1}$	R	(mg/g)	(g/mg.min)		R
Bark	20	11,319	0,165	17,467	12,127	0,028		16,112
				(0,925)				(0,931)

(): value of R

Table 4.38 shows that the pseudo-second order model fits the experimental data slightly better.

In his study, Nogueira (2010) compared the equilibrium times of two activated carbons obtained from Moringa seed husks with a commercial coal for the adsorption of heavy metals. For Pb, she found an equilibrium time of 10 minutes at a pH of 6.5.

Meneghel (2012), in his study of the Pb adsorption process on *Moringa oleifera* cake, found that the model that best fitted the data obtained was the pseudo-second order model, with a correlation coefficient (R^2) equal to 0.999 and the values of q_{exp} and q_{calc} were close.

Casarin (2014), in his study of the adsorption of metal ions using the shell of the Brazil nut seed (Bertholletia excelsa H. B. K), achieved the greatest removal of Pb after 40 minutes, which was the equilibrium time. The pseudo-second order model was the best fit to the experimental data.

4.4.2 Determining the adsorption equilibrium

Based on the tests carried out as described in chapter 3, the concentration of organic matter in the solution at the end of the equilibrium time was measured for different adsorbent masses (c_e) and from this, using equation (2.11), the concentration of organic matter adsorbed by the solid at equilibrium was calculated for the different adsorbent masses (q_e).

Table 4.39 shows the c_e and q_e values for a temperature of 20°C.

Table 4.39: Pb concentration values at equilibrium at 20 °C

c_o 11.5 mg/L; Volume 0.1 L; T 20 °C; 150 rpm; time 60 min

AdsorbentMass (g)	c_e (mg/L)	q_e (mg/g)	% removal
0,02	5,9	28,0	48,7
0,03	5,4	20,4	53,0
0,04	4,8	16,7	58,1
0,05	4,8	13,4	58,0
0,06	4,7	11,3	58,9
0,08	2,8	11,0	76,1
Bark0 ,1	3,5	8,0	69,7
0,15	3,9	5,1	66,5
0,2	2,8	4,4	75,8
0,3	3,2	2,8	72,4
0,4	3,3	2,1	71,2
0,5	2,5	1,8	78,6
0,6	3,0	1,4	74,2

Table 4.39 shows that at 20 °C with a mass of 0.5 g there was the greatest removal of Pb (78.6%).

The Langmuir and Freundlich equilibrium models described in Chapter 2 by equations (2.12) to (2.18) were used to represent the experimental equilibrium data. Figure 4.26 shows the best fit of the Langmuir and Freundlich models to the experimental data. The parameter values of these two equiKbrium models were obtained by fitting the respective non-linear equations to the experimental equiKbrium data, minimising the sum of the squares of the deviations (SQD) between the experimental and calculated values of q_e, using the excel solver.

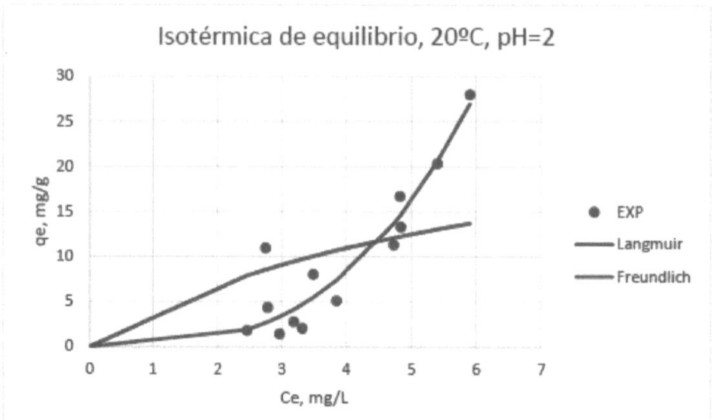

Figure 4.26: Fit of the Langmuir and Freundlich models to the experimental Pb adsorption equilibrium data obtained at 20 °C with moringa bark

The parameter values obtained by the best fit of the theoretical Langmuir and Freundlich models to the experimental data are shown in Table 4.40. This table also shows the correlation coefficient (R) between the experimental values and those calculated by the model.

Table 4.40: Parameter values obtained by the best fit of the Langmuir and Freundlich equiKbrium models to experimental Pb adsorption data using moringa bark

		Modelo de equilíbrio					
		Langmuir			Freundlich		
Adsorve nte	T ºC	q_{max} (mg/g)	K_L (L/mg)	SQD R	K_F (mg/g)/(mg/L)^(1/n)	n^{-1}	SQD R
Casca	20	29,333	0,149	534,7 (0,857)	0,123	3,04	108,8 (0,932)

Table 4.40 and Figure 4.26 show that the experimental equilibrium data is best represented by the Freundlich model with a correlation coefficient (R) of 0.93 between experimental and model-calculated values, although the 1/n parameter of this model is outside the range (between 0 and 1) reported in the literature.

These preliminary lead adsorption results show that it is possible to remove this dissolved metal from wastewater using the same conditions under which soluble organic matter is removed by adsorption.

In their study using Moringa leaves as a biosorbent for lead removal, Reddy *et al.* (2010a) obtained the best fit with the Langmuir model for the experimental data obtained.

Casarin (2014), for the adsorption of Pb on activated charcoal, found removal percentages of up to 100%. The model that best fitted the Pb adsorption process was the Langmuir model.

5 CONCLUSIONS

This work was carried out with the general aim of evaluating the possibility of using Moringa oleifera Lam seeds and pods, from Angola, in the treatment of domestic sewage mixed with industrial effluent, namely as a coagulant and adsorbent.

Samples of raw wastewater (domestic sewage mixed with industrial effluent) were taken from the ZEE wastewater treatment plant in Luanda and characterised at LESRA. This characterisation revealed high turbidity and levels of total suspended solids, organic matter and electrical conductivity values exceeding the emission limits imposed by Angolan legislation. In general, the levels of metals analysed (Pb, Zn, Cu, Cr, Fe, Ni) by atomic absorption spectrophotometry were lower than the detection limit for this technique.

Coagulation-flocculation tests with these raw wastewater samples were carried out using four different coagulants or ways of obtaining the active compound with coagulating properties from the *Moringa oleifera* Lam seed: ground seed with shell, ground seed without shell and aqueous and saline extracts of the ground seed without shell. The coagulation-flocculation conditions that lead to the greatest removal of turbidity were found by varying the initial pH of the sample and the dose of coagulant. For all the coagulants, it was found that pH is a crucial parameter for achieving greater turbidity removal, with the optimum pH of 12 being found for all coagulants. Comparing the four coagulants, it can be concluded that the best coagulant tested is the shelled ground seed because of its turbidity removal power (97%) and COD removal power (62%) at a dose of 120 mg/L lower than that of the other coagulants. It should be noted that a disadvantage of raising the initial pH of the sample to 12 is the increase in its conductivity due to the reagent used to adjust its pH (NaOH).

Adsorption tests of organic matter in the soluble phase were carried out using the seed husk and pods of *Moringa oleifera* Lam which had been pre-treated to increase their adsorbent capacity. The raw wastewater samples were brought to a pH of 12 to sediment most of the solid phase before the adsorption tests. For each of the adsorbents, their zero charge potential or pH value for zero surface charge was determined, being 7.5 for the seed husk and 7.2 for the pod. Adsorption kinetics and equilibrium were studied by fitting theoretical models to experimental data obtained at two different temperatures. For the adsorption of organic matter, it was concluded that the adsorption kinetics are best represented by the pseudo-first order model and the adsorption equilibrium is best represented by the Freundlich model for both adsorbents. The results obtained indicate that the best adsorbent tested is seed husk because it has the highest adsorption capacity.

Lead adsorption tests in the soluble phase were carried out using only pre-treated seed husk. From these tests it can be concluded that the adsorption kinetics are best represented by the pseudo-second order model and the adsorption equilibrium is best represented by the Freundlich model.

Finally, it is concluded that on a laboratory scale it is feasible to use parts of the *Moringa oleifera* Lam tree, ground seed without husk as a coagulant and seed husk as an adsorbent, in the treatment of domestic sewage mixed with industrial effluent by implementing an alternative process consisting of a coagulation-flocculation stage followed by an adsorption stage to replace biological treatment.

For implementation on a WWTP scale, a prior economic evaluation of the alternative process proposed to replace biological treatment is recommended, taking into account the extreme pH values at which the cogulation-flocculation and adsorption stages are carried out.

BIBLIOGRAPHICAL REFERENCES

Akhtar, M., Hasany, M. S., Bhanger, I. M., & Iqbal, S. (2007). Sorption potential of Moringa oleifera pods for removal of organic pollutants from aqueous solutions. *Hazardous materials*, 546-556.

Almeida, M. S. (2018). *Moringa oleifera Lam., its medicinal and nutritional benefits and toxicity assessment.* Master's thesis, University of Coimbra, Pharmaceutical Sciences, Coimbra.

Amaral, L. A., Rossi, O. D., Barros, L. S., lorenzon, C. S., & Nunes, A. P. (2006). Alternative water treatment using moringa oleifera seed extract and solar radiation. *Arq. Instituto Biologico, 73*, 287-293.

Metals handout (20 May 2009). Retrieved 25 May 2009, from radioisotopes.ufrj.br: http://www.radioisotopos.ufrj.br/radioiso/principal.htm.

Araujo, C. S. (2009). *Development of an analytical methodology for the extraction and pre-concentration of Ag (I) using Moringa oleifera Lam.*

Bhatti, H., Mumtaz, B., & et, a. (2007). Removal of Zn (II) ions from aqueous solution using *Moringa oleifera* Lam (horseradish tree) biomass. *Process Biochemistry, 42,* 547-553.

Boaventura, R. F. (1997). *Industrial wastewater treatment.* Notes on water and wastewater treatment.

Boeriu, C. G., Bravo, D., Gosselink, R. A., & Van Dam, J. E. (2004). Characterisation of structure dependent functional properties of lignin with infrared spectroscopy. *Industrial Crops and Products, 20*, 205-218.

Casarin, J. (2014). *Adsorption of metallic tones using the shell of the Brazil nut seed (Bertholletia excelsa H.B.K.) as a biosorbent.* Master's thesis, State University of Western Paraná, Postgraduate Programme in Agronomy, Marechal Candido Rondon.

Chagas, R. C., Saraiva, C. B., Moreira, D. A., SilvaD, J. P., Matos, A. T., & Farage, J. A. (2009). Use of Moringa extract as a coagulating agent in the treatment of dairy wastewater. *National Dairy Congress* , p. 26.

Correa, P. (1984). Dicionario das plantas uteis de Brasil e das exoticas cultivadas. *V*, 276-283.

Cysne, J. R. (2006). *In vitro propagation of Moringa oleifera Lam.* Master's dissertation, Fortaleza.

Presidential Decree No. 261/11 (6 October 2011). *Decree on water quality.* Luanda, Government of Angola, Angola.

Dias, N. P. (2001). Cadmium adsorption isotherms in acric soils. *Brazilian Journal of Agricultural and Environmental Engineering, 5.*

Dobrovol'skii, V. V. (2006). Humic and water migration of heavy metals, Eurasian Soil. *Science, 39*, 1183-1189.

Dotto, G. L., Vieira, M. I., Gongalves, J. O., & Pinto, L. A. (24 February 2011). *Removal of brilliant blue, crepuscular yellow and tartrazine yellow dyes from aqueous solutions using activated carbon, activated earth, diatomaceous earth, quintine and chitosan: Equilibrium and thermodynamic study.* Retrieved 5 November 2005

Duarte, R., & Pasqual, A. (2000). Evaluation of cadmium (Cd), lead (Pb), nickel (Ni) and zinc (Zn) in soils, plants and human hair. *Energy in Agriculture, 15*, 46-58.

Eaton, e. a. (2005). https://www.scrip.org.

ECOSAN. (2012). Wastewater treatment system Viana Industrial Centre.

Febrianto, J., Koshasih, A. N., Sunarso, J., Ju, Y. H., Indraswati, N., & Ismadji, S. (2009). Equilibrium and kinetic studies in adsorption of heavy metal using biosorbent: A summary of recent studies. *Journal of Hazardous Materials*, 616645.

Foidl, N., Makkar, H. P., & Becker, K. (October-November 2001). Potential of Moringa oleifera in agriculture and industry. Dar es salaam, Tanzania.

Franco, M. (2010). *Use of a coagulant extracted from Moringa oleifera seeds as an aid in water treatment by multi-stage filtration.* Master's thesis, State University of Campinas.

Gade, W. K., Buchberger, S. G., Wendell, D., & Kupferle, M. J. (2017). Application of Moringa oleifera seed extract to treat coffee fermentation wastewater. *Hazard Mater Journal, 17*, 102-109.

Gamez, L. L., delRisco, M. L., & Cano, R. E. (2015). Comparative study between M. oleifera and aluminium sulfate for water treatment: case study Colombia. *Environmental Monitoring and Assessment.*

Gassen, M., Gassenschmidt, U., Jany, K. D., Tauscher, B., & Wolf, S. (1990). modern methods in protein and nucleic acid analysis. *Biology Chemitry*, 768-769.

Ghebremichael, K., Gunaratna, K. R., Henriksson, H., H, B., & G., D. (2005). A simple purification and activity assay of the coagulant protein from Moringa oleifera seed. *Water research, 39,* 2338-2344.

Giordano, G. (19 March 2020). *Treatment and control of industrial effluents.* Retrieved from updates@academia-mail.com.

Gueyard, A. J., Barillari, B. R., Lori, B. S., Palmieri, B., & Rollina, P. (11 July 2000). First synthesis of an O-glycosylated isolated from Moringa oleifera. *instituto Sperimentale per le colture industriali.*

Gupta, R., Dubey, D. K., Kannan, G. N., & Flora, S. S. (2007). Concomitant administration of moringa oleifera seed powder in the remediation of arsenic- induced oxidative stress in mouse. *cell biology International 31*, 44-56.

Gupta, V. K., Agarwal, S., Singh, P., & Pathania, D. (15 October 2013). Acrylic acid grafted cellulosic Luffa cylindrical fibre for the removal of dye and metal ions. *Carbohydrate Polymers*, 1214-1221.

Gurgel, L. V., Junior, O. K., Gil, R. P., & Gil, L. F. (31 May 2007). Adsoption of Cu (II), Cd(II), and Pb (II) from aqueous single metal solutions by cellulose and mercerised celluse chemically modified with succinic anhydride. *ScienceDirect*, 3077-3083.

Heredia, J., & Martin, J. (2008). Removing heavy metals from polluted surface water with a tannin-based flocculant agent. *Hazardous Materials.*

Jahn, S. A. (June 1988). Using Moringa seeds as coagulants in developing countries. *The American Water Works Association*, 43-50.

Katayon, s., Noor, M. M., Asma, M., Ghani, I. A., Thamer, A. M., & Azni, e. a. (2006). Effects of storage conditions of moringa oleifera seeds on its performance in coagulation. *Bioresource Technology, 97 (13)*, 1455-1460.

Kreush, M. A. (2005). *Evaluation with proposals for improvement of the industrial lead recycling process and indication of applicability for the slag generated.* Master's thesis, Federal University of Parana, Parana.

Kumari, P., Sharma, P., Srivastava, P., & Srivastava, M. M. (2005). Biosorption studies on shelled moringa oleifera Lam seed powder: Removal and recovery of arsenic from aqueous system. *International Journal of Mineral Processing, 78,* 131-139.

Kunz, A., Zamora, P. P., Moras, S. G., & Duran, N. (2002). *New trends in textile effluent treatment.* Federal University of Paraná, Chemistry, Curitiba.

Leitao, A. A. (2018). Introduction to water and wastewater treatment.

Leitao, A., Jose, F., Francisco, T., Chipate, A., Serrao, R., & Gongalves, J. (2017). Determination of chemical oxygen demand in samples considered difficult to analyse.

Lenardao, E. J., Freitag, R. A., Dabdoub, M. J., Batista, A. C., & Silveira, C. C. (2003). The 12 principles of green chemistry and their application to teaching and research activities. *Qumica Nova,* 123-129.

Madrona, G. S., Serpelloni, G. B., Vieira, A. M., Nishi, L., Cardoso, K. C., & Bergamasco, R. (2010). Study of the effect of saline solution on the extraction of the Moringa oleifera seeds active component for water treatment. *Water, Air & Soil Pollution,* 409-415.

Mangale, S. M., Chonde, S. G., & Raut, P. D. (March 2012). Use of Moringa Oleifera (Drumstick) seed as natural absorbent and an antimicrobial agent for ground water treatment. *Research Journal of recent Sciences, 1 (3),* 31-40.

Martins, A. A., Oliveira, R. M., & Guarda, E. A. (2014). Potential use of organic compounds as coagulants, flocculants and adsorbents in water and effluent treatment. *Environmental Forum of Alta Paulista, 10.*

Mataka, L. M., Henry, E. T., Masamba, W. L., & Sajidu, S. M. (2006). Lead remediation of contaminated water using Moringa stenopetala and Moringa oleifera seed powder. *Environment Science technology, 3,* 131-139.

McCabe, W. L., Smith, J. C., & Harriot, P. (1993). Units operations of chemical engeering. *Mc Graw Hill, 5ª ed,* 810-821.

Melo, C. R. (2009). *Synthesis of Zeolite type 5A from kaolin for the adsorption of heavy metals from aqueous solutions.* Master's thesis, Federal University of Santa Catarina, Department of Chemical Engineering, Florianopolis.

Mendes, F., & Coelho, N. (2007). Study of the use of Moringa oleifera to remove silver and manganese from water. *Horinzonte Cientifico.*

Meneghel, A. P. (2012). *Remediation of water contaminated by metals (Cd, Pb and Cr) using Moringa oleifera Lam seed cake as an adsorbent.* Master's Degree Report, State University of Western Paraná, Postgraduate Programme in Agronomy, Marechal Candido Rondon.

Metcalf, &, & Eddy. (1991). *Wastewater Engineering: treatment, Disposal, Reuse* (Vol. 3rd ed). New York, USA.

Monaco, L., Vieira, P. A., Antonio, T. M., Cello, A. R., Filipe, S. N., & Antover, P. S. (2010). Validation of moringa seed extract as a coagulating agent in the treatment of water supply and wastewater. *Ambiente e agua, 5*(3), 222-231.

Monte, H. M., Santos, M. T., Barreiros, A. M., & Albuquerque, A. (2016). *Wastewater treatment "Physical and chemical treatment operations and processes"* (Vol. 5). Lisbon, Portugal: Entidade Reguladora dos Servigos de Aguas e Res^duos (ISPEL).

Monteiro, R. A. (2009). *Avaliação do potencial de adsorgao de U, Th, Pb, Zn e Ni pelas fibras de coco.* dissertação de Mestrado, Universidade de Sao Paulo, Instituto de pesquisa energética e nuclear.

Morais, F. M. (2013). *Study of a new urban wastewater treatment process aimed at minimising the operational difficulties of the activated sludge process and its applicability in countries lacking in energy and technology.* Master's dissertation, University of Porto, Porto.

Muyibi, S. A., & Evison, L. M. (December 1995). Optimising physical parameter affecting coagulation of turbid water with Moringa oleifera seeds. *Water Research, 29*(12), 2689-2695.

Nascimento, R. F., Lima, A. C., Vidal, C. B., Melo, D. Q., & Raulino, G. S. (2014). Adsorption: Theoretical aspects and environmental applications.

Ndabigengesere, A., Narasiah, K. S., & Talbot, B. G. (February 1995). Active agents and mechanism of turbid waters using Moringa oleifera. *Water Research, 29*(2), 706-710.

Ndbigengesere, A., & Narasiah, K. S. (1998). Quality of water treated by coagulation using Moringa oleifera seed. *Water resources, 32*(3), 781-791.

Ndibewu, P. P., Mnisi, R. L., Mokgalaka, S. N., & Mccrindle, R. I. (2011). Heavy metal removal in aqueous systems using Moringa oleifera. *Materials Science and Engineering B*, 843-853.

Nogueira, M. W. (2010). *The use of activated charcoal produced from Moringa oleifera bark as an adsorbent in the removal of heavy metals from water.* Master's dissertation, Federal University of Ouro Preto, Institute of Exact and Biological Sciences, Ouro Preto.

Okuda, T., Baes, A. U., Nishijima, W., & Okada, M. (1999). Improvement of extraction method of coagulation active components from Moringa oleifera seed. *Water Reserch, 33,* 3373-3378.

Okuda, T., Baes, A. U., Nishijima, W.,   & Okado, M. (2001). Isolation and characterisation of coagulent extracted from Moringa oleifera seed by salt solution. *Water research, 35 (2)*, 405-410.

Pagnanelli, F., Mainelli, S., Veglio, F., & Toro, L. (2003). Heavy metal removal by olive pomace: biosorbent characterization and equilibrium modelling. *Chemestry Enginering, 58,* 4709-4717.

Pavanelli, G. (2001). *Effectiveness of different types of coagulants in the sedimentation of water with high colour or turbidity.* Master's thesis, University of Sao Paulo, Sao Carlos School of Engineering.

Pereira, D. P., Araujo, N. A., Santos, T. M., Santana, C. R., & Silva, G. F. (2011). Utilisation of the Moringa oleifera Lam torat to treat produced water. *9 n. 3,* 323-331.

Perpetuo, E. A. (2020). *Parameters for characterising the quality of industrial water and effluents.* Retrieved 03 September 2020, from updates@academia-mail.com

Perreira, V. F. (2008). *Removal of zinc (II) ions from effluents derived from electroplating processes using chemically modified vegetable fibre waste.* Master's dissertation, Federal University of Ouro Preto, Exact and Biological Sciences.

Petroni, L. G., & Pires, M. F. (2000). Adsorption of zinc and cadmium in peat columns. *Qumica Nova, 23,* 477-481.

Price, M. L., & Davis, K. (2000). *The Moringa tree. Echo technical note.* Retrieved 25 June 2009, from Echonet.org: http://www.echonet.org/

Reddy, D. H., Harinath, Y., Seshaiah, K., & Reddy, A. V. (15 August 2010 b). Biosorption of Pb (II) from aqueous solutions using chemical modified Moringa oleifera tree leaves. *Chemical Engineering, 162,* 626-634.

Reddy, D. H., Ramana, D. K., Seshaiah, K., & Reddy, A. V. (1 March 2011). Biosorption of Ni (II) from aqueous phase by Moringa oleifera bark, a low cost biosorbent. *Desalination, 268,* 150-157.

Reddy, D. H., Seshaiah, K., Reddy, A. V., Rao, M. M., & Wang, M. C. (15 February 2010 a). Biosorption of Pb 2+ from aqueous solutions by Moringa oleifera bark: equilibrium and kinetic studies. *Journal of Hazardous Materials, 174,* 831838.

Regalbuto, J. R.,   & Robles, J. (2004). The engineering of Pt/Carbon catalysts preparation. *Progress Report.*

Ribeiro, A. T. (2010). *Application of Moringa oleifera in the treatment of water for human consumption.*

Rico, T. E., Santos, L. M., Reis, E. M., Silva, F. F., & Zonetti, P. C. (July-December 2010). Treatment of tannery wastewater using moringa seeds (Moringa oleifera Lam). *Agroambiente, 4,* 96-101.

Romanielo, L. L. (1999). *Mathematical and thermodynamic modelling of multicomponent gas adsorption.* PhD Thesis, State University of Campinas, Campinas.

Rosa, D. (April 1993). *Moringa oleifera: A Perfect Tree for Home Gardens.* (The agroforestry Information Service 1010) Retrieved April 1993, from www.winrock.org/forestry/factnet.htm.

Santana, C. R. (n.d.). *Treatment of produced water through the flotation process using Moringa oleifera Lam as a natural coagulant.*

Santos, A. F. (2012). *Study of the water quality of Luanda Bay: distribution of heavy metals in water, suspended solids and sediments.* PhD thesis.

Santos, X. A. (2016). *Production of activated carbon from Moringa oleifera seed husk for use as an adsorbent in the removal of ibuprofen from effluents.* Master's degree report, Federal University of Sao Joao Del-Rei, Alto Paraopeba Campus.

Schmitt, D. M. (2011). *Treatment of dairy industry wastewater by combined coagulation/flocculation/ adsorption/ultrafiltration processes using Moringa oleifera seed as a coagulant.* Western Paraná State University, Centre for Engineering and Exact Sciences, Toledo.

Schwanke, R. O. (2003). *Determination of the diffusivity of aromatic hydrocarbons in Zeolites y by chromatographic methods.* Master's thesis, Federal University of Santa Catarina, Chemical Engineering and Food Engineering, Florianopolis.

Sharma, P., Kumari, P., Srivastav, M. M., & Srivastava, S. (2006). Removal of cadmium from aqueous system by shelled Moringa oleifera Lam. seed powder. *Bioresource Technology, 97,* 299-305.

Sharma, P., Kumari, P., Srivastava, M. M., & Srivastava, S. (2007). Ternary biosorption studies of Cd (II), Cr (III) and Ni (II) on shelled moringa oleifera seeds. *Bioresource Technology, 98,* 474-477.

Silva, C. A. (2005). *Studies applied to the use of Moringa oleifera as a natural coagulant for improving water quality.* Master's dissertation, Federal University of Uberlandia.

Silva, E. (2009). *Optimisation of the coagulation-flocculation stage in a water treatment plant using a natural coagulant.* Agostino neto University.

Silva, F. J., & Matos, J. E. (2010). On the dispersion of Moringa oleifera for water treatment. *Tecnologia Fortaleza, 29*(2), 157-163.

Silva, G. F., Santana, M. F., Lima, A. K., Bergamasco, R., Paiva, P. M., Sant'anna, M. C., . . . Bery, C. C. (2018). The *potential of Moringa oleifera Lam* (Vol. IV). Sao Cristovao, Brazil: UFS.

Silveira, G. E. (2010). *Industrial effluent treatment system.* Master's dissertation, UFRGS.

Souza, H. K. (2016). *Use of the seed, bark and pod of Moringa oleifer Lam in the biosorption process to remove Diuron from contaminated water.* Master's thesis, Maringa.

Stein, R. T. (2012). *Characterisation and evaluation of the effluent treatment system of a food industry, with a view to reuse.*

Trindade, T., & Manuel, R. (2006). *Treatability tests on wastewater (Physico-chemical treatment: coagulation/flocculation).*

Vieira, A. M., Vieira, M. F., Silva, G. F., Araujo, A. A., Klen, M. R., Veit, M. T., & Bergamasco, R. (5 June 2009). Use of Moringa oleifera seed as a natural adsorbent for wastewater treatment. *I Water, Air and soil Pollution*, 273281.

Von Sperling, M. (1996). *Introduction to water quality and sewage treatment* (Vol. 2). (S. (031)411-7077, Ed.) Belo Horizonte, Minas Gerais, Brazil.

Zeng, G., Jiang, R., Guohe, H., Xu, M., & Li, J. (January 2007). Optimisation of wastewater treatment alternative selection by hierarchy grey relational analysis. *Environmental Management, 82,* 250-25

ANNEX

This annex shows the flowchart of the treatment system adopted at the WWTP in the EEZ.

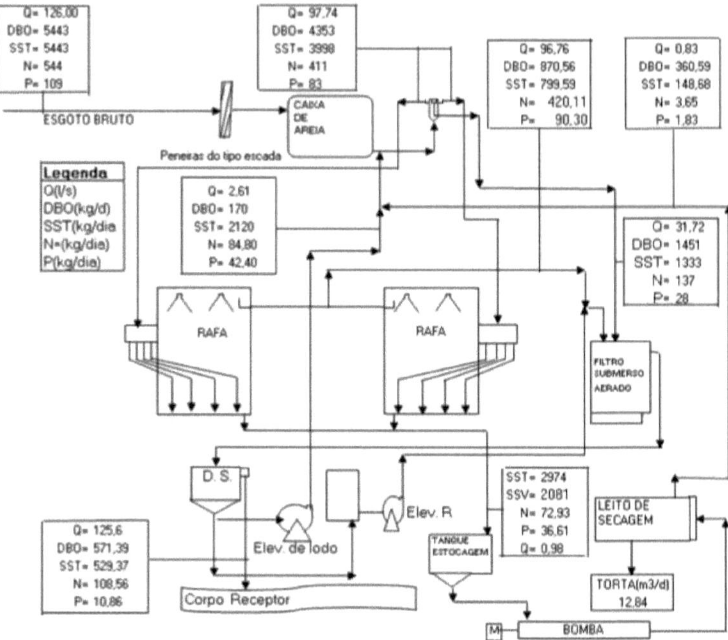

Source: ECOSAN (2012)

yes
I want morebooks!

Buy your books fast and straightforward online - at one of world's fastest growing online book stores! Environmentally sound due to Print-on-Demand technologies.

Buy your books online at
www.morebooks.shop

Kaufen Sie Ihre Bücher schnell und unkompliziert online – auf einer der am schnellsten wachsenden Buchhandelsplattformen weltweit! Dank Print-On-Demand umwelt- und ressourcenschonend produzi ert.

Bücher schneller online kaufen
www.morebooks.shop

Printed by Books on Demand GmbH, Norderstedt / Germany